MW01141742

Monographs in Electrical and Electronic Engineering 17

General Editors: R. L. Grimsdale, P. Hammond,
and E. H. Rhoderick

Monographs in Electrical and Electronic Engineering

10. *The theory of linear induction machinery* (1980) Michel Poloujadoff
11. *Metal–semiconductor contacts* (1978, reissued 1980) E. H. Rhoderick
12. *Energy methods in electromagnetism* (1981) P. Hammond
13. *Low-noise electrical motors* (1981) S. J. Yang
14. *Large synchronous machines* (1981) J. H. Walker
15. *Superconducting rotating electrical machines* (1983) J. R. Bumby
16. *Stepping motors and their microprocessor controls* (1984) T. Kenjo
17. *Machinery noise measurement* (1985) S. J. Yang and A. J. Ellison
18. *Permanent-magnet and brushless DC motors* (1985) T. Kenjo and S. Nagamori

Machinery Noise Measurement

S. J. Yang

Heriot–Watt University, Edinburgh

and

A. J. Ellison

The City University, London

CLARENDON PRESS · OXFORD · 1985

Oxford University Press, Walton Street, Oxford OX2 6DP
Oxford New York Toronto
Delhi Bombay Calcutta Madras Karachi
Kuala Lumpur Singapore Hong Kong Tokyo
Nairobi Dar es Salaam Cape Town
Melbourne Auckland
and associated companies in
Beirut Berlin Ibadan Nicosia

Oxford is a trade mark of Oxford University Press

Published in the United States
by Oxford University Press, New York

British Library Cataloguing in Publication Data
Yang, S. J.
Machinery Noise Measurement.—(Monographs in electrical and electronic engineering)
1. Noise control 2. Machinery—Noise
I. Title II. Ellison, A. J. III. Series
620.2'3 TJ179

ISBN 0-19-859333-3

Library of Congress Cataloging in Publication Data
Yang, S. J.
Machinery Noise Measurement.
(Monographs in electrical and electronic engineering)
Bibliography: p.
Includes index.
1. Machinery—Noise—Measurement. I. Ellison, A. J.
II. Title. III. Series.
TJ179.Y35 1985 621.8'11 84-20743
ISBN 0-19-859333-3

Typeset and printed in Northern Ireland by
The Universities Press (Belfast) Ltd.

Preface

This is a general book on machinery noise measurement for engineers. It is intended to be of value to the designers and manufacturers of any machinery or equipment which makes noise—for example, electric and mechanical machinery, office and data-processing equipment, and domestic appliances—with the exception of moving vehicles.

The noise level produced by engineering equipment is growing in importance with increasing emphasis on the reduction of 'noise pollution' as part of governments' efforts to improve the 'quality of life'. Home and overseas purchasers of engineering equipment demand quietness and frequently compare the products of different manufacturers for relative noise production. However, it is often the case that such comparisons are meaningless because they are not based on the same measurement methods. An important indicator of high quality, a vital factor if international trade is to grow, is a careful and valid specification of noise produced. An engineering exporter who does not pay proper attention to the noise of his equipment and cover this feature in his literature and technical specifications does so at his peril.

There is no dearth of bulky scientific books on acoustics. But most seem to have been written for specialists and do not provide much help to an engineer having a specific noise problem to solve. There has been a real need for an up-to-date review of advances in our knowledge of noise and its measurement over the last twenty years, with particular reference to engineering matters. Only a very unaware engineering designer will not have given thought to the need for quietness in his firm's products and appreciated how important it is, if noise figures are to be taken seriously, for the measurements on which they are based to have been made rationally and accurately. An engineer, producing a product for sale against competition, needs advice on acquiring microphones and noise equipment, on how to use it, on how to specify appropriate measurements, and how to deal with the results. And the problem of making measurements of value when no anechoic chamber is available is a common one. It is often not appreciated that perfectly adequate noise measurements can often be made without such an expensive facility.

The authors' intention in producing this book has been to provide soundly based advice for their fellow engineers who are not noise experts, but who have the need to measure and evaluate noise from machinery. We provide this advice in practical form, with worked examples. With international trade in mind we also append a list of national and international standards on machinery noise.

Recently, advances have been made in digital techniques. We therefore cover digital sound-level meters and spectrum analysers in addition to giving a comprehensive cover to the more conventional analogue instruments.

Edinburgh and London
August 1984

S. J. Y.
A. J. E.

Acknowledgements

We should like to thank the Science and Engineering Research Council of the United Kingdom for supporting financially the investigation of machinery noise problems at Heriot–Watt University, Edinburgh and Queen Mary College, London over a period of twenty years since the early 1960s. This book contains some of the results of that investigation made by us in association with Dr C. J. Moore.

We are very much indebted to the Institution of Electrical Engineers, London; the Institute of Electrical and Electronic Engineers, New York; the Institute of Noise Control Engineering, New York; the International Electrotechnical Commission, Geneva; the British Electrical and Allied Manufacturers' Association, London; American Institute of Physics, New York; the British Standards Institution, London; the International Organisation for Standardization, Geneva; Brüel & Kjaer, Naerum; Academic Press Inc., London; S. Hirzel Verlag, Stuttgart; and the Institute of Petroleum, London, for permission to use the material in many papers and/or standards published by them.

Grateful thanks must go to Mr. R. Jensen for his constructive comments on the draft and to Mrs. F. Samson, Mrs. P. Ingram and Miss P. Meikle for typing the manuscript.

Special thanks are due to Professor R. Hanitsch, Technical University of Berlin; Dr. P. L. Timar, Technical University of Budapest and Mr. J. D. Bradley, Heriot–Watt University, who have translated many international and national standards into English for this book. Finally, the editorial and other assistance provided by Oxford University Press is gratefully acknowledged.

Edinburgh and
London
November 1984

S. J. Y.
A. J. E.

TO FEI, ELLEN
MARIAN, JENNIFER, AND RICHARD

Contents

1. Fundamental terminology

A steady pure-tone sound in air is a sound which exhibits sinusoidal pressure variations in air having a constant frequency and a constant amplitude. The noise, i.e. unwanted sound, around us is usually a combination of a series of components at different frequencies of various mechanical, electromagnetic, and aerodynamic origins. Before examining noise-measurement problems, we shall first introduce some fundamental terminology in the field of acoustic noise.

1.1. Sound-pressure level

In a sound field the sound pressure at a given point is the instantaneous pressure minus the static pressure at that point. The sound pressure is generally expressed as the sound-pressure level in dB (decibel) defined by

$$L_p = 10 \log_{10}\left(\frac{p^2}{p_{ref}^2}\right) \text{ dB} \tag{1.1}$$

where p = the r.m.s. sound pressure and p_{ref} = the r.m.s. reference sound pressure.

The reference sound pressure is internationally taken as $2 \times 10^{-5}\ \text{N m}^{-2}$ (i.e. 20 μPa), which is approximately equal to the r.m.s. sound pressure of a pure tone of 1000 Hz at the normal threshold of hearing. Almost all noise measuring equipment commercially available gives direct readings in sound-pressure level as defined by eqn (1.1) and based on the reference pressure of $2 \times 10^{-5}\ \text{N m}^{-2}$.

Example 1.1

A pure tone sound exhibits a r.m.s. sound pressure of $2 \times 10^{-3}\ \text{N m}^{-2}$. Calculate the sound-pressure level. The sound-pressure level is

$$L_p = 10 \log_{10}\left(\frac{2 \times 10^{-3}}{2 \times 10^{-5}}\right)^2 = 40 \text{ dB}.$$

1.2. Sound-power level

The sound power emitted by a source is, for ease of comparison, expressed as the sound-power level in dB defined by

$$L_W = 10 \log_{10}\left(\frac{W}{W_{ref}}\right) \tag{1.2}$$

where W = the average sound power emitted by an object in watts and W_{ref} = the reference sound power in watts. The reference sound power W_{ref} is internationally taken as 1×10^{-12} W (i.e. 1 pW). In the older literature, the reference sound power was sometimes taken as 1×10^{-13} W.

Example 1.2

An electric motor emits a sound power of 1×10^{-6} W at 1000 Hz. Determine the sound-power level. The sound-power level is

$$L_W = 10 \log_{10}\left(\frac{1 \times 10^{-6}}{1 \times 10^{-12}}\right) = 60 \text{ dB} \quad \text{re} \quad 1 \times 10^{-12} \text{ W}.$$

1.3. Sound intensity

The sound intensity at a given point in a sound field in a specified direction is defined as the average sound power passing through a unit area perpendicular to the specified direction at that point. For a plane and spherical sound wave propagating in a free field (a field free from reflections) the sound intensity along the direction of wave propagation is given by[1.1]

$$I = \frac{p^2}{\rho c} \text{ W m}^{-2} \tag{1.3}$$

where p = the r.m.s. sound pressure in N m^{-2}, ρ = the constant equilibrium density of the medium in kg m^{-3}, and c = the velocity of sound in the medium in m s^{-1}.

The product of ρ and c is defined as the characteristic impedance of the medium. At 20 °C and standard atmospheric pressure, the characteristic impedance of air is

$$\begin{aligned}(\rho c)_{air} &= (1.21 \text{ kg m}^{-3})(343 \text{ m s}^{-1}) \\ &= 415 \text{ kg m}^{-2} \text{ s}^{-1}.\end{aligned}$$

In acoustics, the unit in kg m^{-2} s^{-1} is also called a rayl. (The c.g.s. rayl is sometimes used also.)

1.4. Relationship between sound-power level and sound-pressure level

Let us consider the case of a machine completely enclosed by a surface. Assuming that the direction of wave propagation at any point on the surface is perpendicular to the surface and that the sound wave can be regarded as either a plane or a spherical wave, the total sound power emitted by the machine is

$$W = \int_A I_i \, dA_i = \int_A \frac{p_i^2}{\rho c} \, dA_i \tag{1.4}$$

where the integration is over the whole surface area A and p_i is the sound pressure on the ith elementary area dA_i. If the whole surface area is divided into n equal parts dA and the value of n is sufficiently large, eqn (1.4) becomes

$$W = \sum_{i=1}^{n} \frac{p_i^2}{\rho c} \, dA$$
$$= n(dA)\left(\frac{1}{n}\sum_{i=1}^{n}\frac{p_i^2}{\rho c}\right) = A\frac{p_{av}^2}{\rho c} \tag{1.5}$$

where p_{av} = the average r.m.s. sound pressure over the whole surface area.

Dividing both sides of eqn (1.5) by W_{ref} and taking logarithms, we have the sound-power level

$$L_W = 10 \log_{10} \frac{W}{W_{ref}} = 10 \log_{10} A + 10 \log_{10} \left(\frac{p_{av}^2}{W_{ref}\rho c}\right). \tag{1.6}$$

Based on the definition of the sound-pressure level, the value of p_{av}^2 can be expressed in terms of the sound-pressure levels on the surface by

$$p_{av}^2 = p_{ref}^2 \left(\frac{1}{n}\sum_{i=1}^{n} 10^{0.1L_{p,i}}\right) \tag{1.7}$$

where $L_{p,i}$ is the sound-pressure level at the ith elementary area.

Combining eqns (1.6) and (1.7),

$$L_W = 10 \log_{10} A + 10 \log_{10}\left(\frac{p_{ref}^2}{W_{ref}\rho c}\right) + 10 \log_{10}\left(\frac{1}{n}\sum_{i=1}^{n} 10^{0.1L_{p,i}}\right). \tag{1.8}$$

The second term of eqn (1.8) is approximately equal to zero when $p_{ref} = 2\times10^{-5}$ N m^{-2}, $W_{ref} = 1\times10^{-12}$ W, and $\rho c = 415$ kg m^{-2} s^{-1}. Thus the sound-power level can be expressed as

$$L_W = 10 \log_{10} A + \bar{L}_p \tag{1.9}$$

where A = the whole surface area enclosing the machine in m^2 and

$$\bar{L}_p = 10 \log_{10}\left(\frac{1}{n}\sum_{i=1}^{n} 10^{0.1L_{p,i}}\right). \tag{1.10}$$

The value of $\bar{L}_p$ is called the *mean sound-pressure level* or the *level of mean-square sound pressure.*

Equation (1.9) gives a simple approximate relationship between the sound-power level and the mean sound-pressure level and enables us to determine the sound-power level from sound-pressure level measurements made over an 'imaginary' surface enclosing the machine. Strictly speaking, eqn (1.9) is valid only when the sound-pressure level measurements are made in a free field, e.g. in an anechoic chamber. However, for many practical cases, eqn (1.9) can be used for measurements made in an ordinary laboratory if suitable corrections are made (see Chapter 4).

1.5. Combining two or more sounds of different frequency

Let us assume that there are two pure tones at frequencies f_1 and f_2 ($f_1 \neq f_2$) in a sound field. Let the r.m.s. sound pressure of these two sounds be p_1 and p_2, respectively. Based on the summation of the two sinusoidal waves, the total r.m.s. sound pressure is

$$p_{\text{tot}} = (p_1^2 + p_2^2)^{1/2}. \tag{1.11}$$

The above equation can be rewritten as

$$\frac{p_{\text{tot}}^2}{p_{\text{ref}}^2} = \frac{p_1^2}{p_{\text{ref}}^2} + \frac{p_2^2}{p_{\text{ref}}^2}.$$

Taking logarithms and multiplying both sides by 10, we have

$$10 \log_{10} \frac{p_{\text{tot}}^2}{p_{\text{ref}}^2} = 10 \log_{10} \left(\frac{p_1^2}{p_{\text{ref}}^2} + \frac{p_2^2}{p_{\text{ref}}^2} \right),$$

i.e.

$$L_{\text{p,tot}} = 10 \log_{10}(10^{0.1L_{\text{p},1}} + 10^{0.1L_{\text{p},2}}). \tag{1.12}$$

If the two pure tones have the same r.m.s. sound pressure, hence the same sound-pressure level, the combined total sound-pressure level is, based on eqn (1.12), only 3 dB more than the sound-pressure level of one pure tone alone.

The above procedure can be generalized to include any number of pure tones, each of which has a frequency different from the rest. The combined sound-pressure level of n components of different frequencies is

$$\begin{aligned} L_{\text{p,tot}} &= 10 \log_{10} \left(\frac{p_1^2}{p_{\text{ref}}^2} + \frac{p_2^2}{p_{\text{ref}}^2} + \cdots + \frac{p_n^2}{p_{\text{ref}}^2} \right) \\ &= 10 \log_{10}(10^{0.1L_{\text{p},1}} + 10^{0.1L_{\text{p},2}} + \cdots + 10^{0.1L_{\text{p},n}}). \end{aligned} \tag{1.13}$$

It should be emphasized that the combination of two or more sound waves of different frequencies does not depend on the phase angle of the sound pressure of these waves. However, the combination of two sound waves of the same frequency does depend on the phase angle. Let us consider the case of two sound waves of the same frequency in a noise field. The combined total r.m.s. sound pressure at a given point in the field is, based on the addition of two phasors,

$$p_{tot}=(p_1^2+p_2^2+2p_1p_2\cos\theta)^{1/2} \tag{1.14}$$

where p_1 and p_2 are the r.m.s. sound pressures of the two sound waves, and θ is the phase angle between the two sound waves at a given point. If $p_1=p_2$ and the two sound waves are in antiphase at a particular point, i.e. $\theta=180°$, then the total sound pressure becomes zero at that point. On the other hand, if the two sound waves are in phase with each other at another point, i.e. $\theta=0$, then the total sound pressure at that point is doubled, resulting in an increase of 6 dB in the sound-pressure level.

1.6. Octave- and one-third octave-band sound-pressure levels

Sound-measuring equipment is usually equipped with a number of band-pass filters and the final reading of the equipment gives the total sound-pressure level in a particular frequency band. The frequency response of a typical octave bandpass filter is shown in Fig. 1.1. The general relationship between the upper cut-off frequency f_2 and the lower cut-off frequency f_1 is

$$f_2=2^a f_1 \tag{1.15}$$

where a is an arbitrary constant. For the most common filters used in

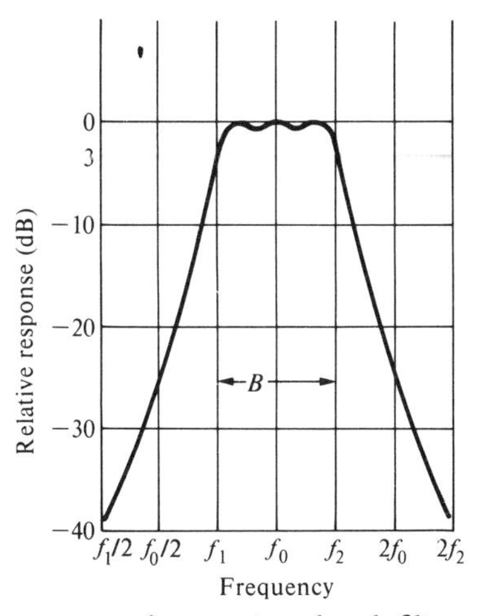

Fig. 1.1. Frequency response of an octave-band filter showing lower cut-off frequency and the upper cut-off frequency. f_1 = lower cut-off frequency; f_0 = centre frequency; f_2 = upper cut-off frequency; B = pass band bandwidth.

noise-measuring equipment, a is 1 or $\frac{1}{3}$. When $a=1$, the filter is called an octave-band filter and when $a=\frac{1}{3}$ the filter is a one-third octave-band filter. The bandwidth of the passband, i.e. the width between two −3 dB points (see Fig. 1.1), is

$$B=f_2-f_1. \tag{1.16}$$

The centre frequency f_0 of a filter is defined as

$$f_0=\sqrt{(f_1 f_2)}. \tag{1.17}$$

TABLE 1.1. Centre and approximate cut-off frequencies for octave- and one-thord octave-band filters[1.3]*

Octave bands			One-third-octave bands		
Centre frequency f_0 (Hz)	Approximate lower cut-off frequency f_1 (Hz)	Approximate upper cut-off frequency f_2 (Hz)	Centre frequency f_0 (Hz)	Approximate lower cut-off frequency f_1 (Hz)	Approximate upper cut-off frequency f_2 (Hz)
16	11	22	16.0	14.1	17.8
			20.0	17.8	22.4
			25.0	22.4	28.2
31.5	22	44	31.5	28.2	35.5
			40.0	35.5	44.7
			50.0	44.7	56.2
63	44	88	63.0	56.2	70.8
			80.0	70.8	89.1
			100.0	89.1	112.0
125	88	177	125.0	112.0	141.0
			160.0	141.0	178.0
			200.0	178.0	224.0
250	177	355	250.0	224.0	282.0
			315.0	282.0	355.0
			400.0	355.0	447.0
500	355	710	500.0	447.0	562.0
			630.0	562.0	708.0
			800.0	708.0	891.0
1 000	710	1 420	1 000.0	891.0	1 122.0
			1 250.0	1 122.0	1 413.0
			1 600.0	1 413.0	1 778.0
2 000	1 420	2 840	2 000.0	1 778.0	2 239.0
			2 500.0	2 239.0	2 818.0
			3 150.0	2 818.0	3 548.0
4 000	2 840	5 680	4 000.0	3 548.0	4 467.0
			5 000.0	4 467.0	5 623.0
			6 300.0	5 623.0	7 079.0
8 000	5 680	11 360	8 000.0	7 079.0	8 913.0
			10 000.0	8 913.0	11 220.0
			12 500.0	11 220.0	14 130.0
16 000	11 360	22 720	16 000.0	14 130.0	17 780.0
			20 000.0	17 780.0	22 390.0

* (Extracts from ISO Standards are reproduced by permission of the British Standards Institution, 2 Park Street, London, from whom complete copies of the publication can be obtained.)

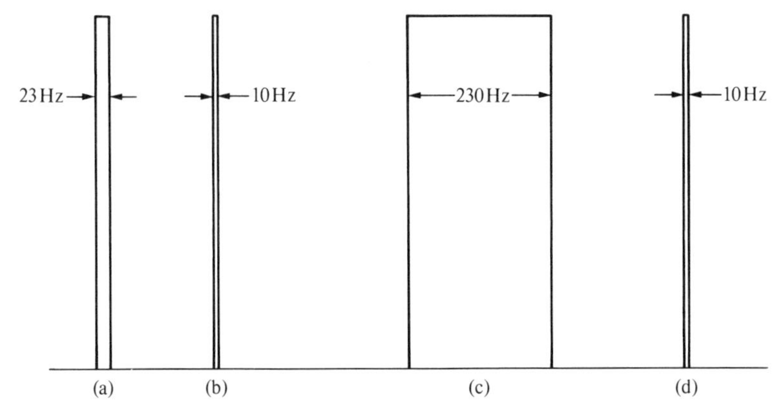

Fig. 1.2. Bandwidths of a constant 23 per cent bandwidth filter and a constant 10-Hz bandwidth filter. (a) Bandwidth of 23 per cent filter at 100 Hz; (b) bandwidth of 10-Hz filter at 100 Hz; (c) bandwidth of 23 per cent filter at 1000 Hz; (d) bandwidth of 10-Hz filter at 1000 Hz.

The centre frequency and frequency band of each of the internationally standardized octave- and one-third octave-band filters are shown in Table 1.1. In addition to octave-bands, and one-third octave bands, some noise analysers used for identifying the predominant frequency components of a noise are equipped with narrow-band filters. The bandwidth of narrow-band filters is either expressed as a constant percentage of the tuned-in centre frequency or as a fixed value in hertz, regardless of the tuned-in centre-frequency value. These are described as constant-percentage bandwidth filters and constant bandwidth filters, respectively. The octave- and one-third octave-band filters are constant 71 per cent and 23 per cent bandwidth filters respectively. Figure 1.2 shows the bandwidths of a one-third octave-band filter at 100 and 1000 Hz, compared with that of a 10 Hz constant bandwidth filter.

If the narrow-band sound-pressure levels of all important frequency components within the passband of a one-third octave-band are known, the corresponding one-third octave-band sound-pressure level can be found by using eqn (1.13). Similarly, one-third octave-band sound-pressure levels can be combined to give the octave-band sound-pressure level (see Ex. 1.3).

Example 1.3

The one-third octave-band sound-pressure levels with centre frequencies of 50, 63, and 80 Hz are 27.9, 24.8, and 23.3 dB, respectively. Calculate the octave-band sound-pressure level with a centre frequency of 63 Hz. Based on eqn (1.13), the required octave-band sound-pressure level is

$$L_p = 10 \log_{10}(10^{2.79} + 10^{2.48} + 10^{2.23}) = 30.5 \text{ dB}.$$

1.7. A-weighted sound level

Based on many tests of people with normal hearing in the age group of 18 to 25 years, a set of normal equal loudness contours as shown in Fig. 1.3 has been established. These contours show that the sensitivity of the ear varies with frequency and pressure level. In order to simulate very roughly the variation of the ear sensitivity with frequency, a frequency weighting called A-weighting has been standardized by the International Electrotechnical Commission.

The frequency response of A-weighting (see Fig. 1.4) takes a frequency of 1000 Hz as the reference and gives predetermined positive or negative adjustments to all other frequencies in such a way as to match roughly the Fletcher–Munson equal-loudness contour[1.4] passing through 30 dB at 1000 Hz. For example, it gives an approximate 19–dB attenuation at 100 Hz since the human ear regards a 100-Hz pure tone having a sound-pressure level of 29 dB as equally loud as a 1000-Hz pure tone having a sound-pressure level of 10 dB (see Fig. 1.3).

The noise meter reading obtained using the A-weighting network, is called the A-weighted sound level (or simply the sound level A) with a unit of dB(A) (or simply dBA). The A-weighted sound level represents the combined total sound-pressure level in the entire audible frequency range of 20 Hz to 20 kHz with A-weighting adjustments. From eqn (1.13), the A-weighted sound level of a noise is given by

$$L_{\mathrm{A}} = 10 \log_{10}\left(\frac{p^2_{\mathrm{tot,A}}}{p^2_{\mathrm{ref}}}\right) = 10 \log_{10}\left(\frac{p^2_{1,\mathrm{A}}}{p^2_{\mathrm{ref}}} + \frac{p^2_{2,\mathrm{A}}}{p^2_{\mathrm{ref}}} + \cdots + \frac{p^2_{m,\mathrm{A}}}{p^2_{\mathrm{ref}}}\right)$$

$$= 10 \log_{10} \sum_{i=1}^{m} 10^{0.1L_{\mathrm{p,i,A}}} \qquad (1.18)$$

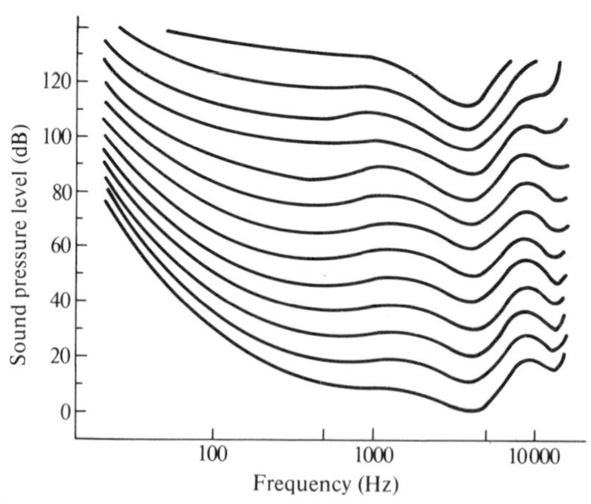

Fig. 1.3. Equal loudness contours for pure tones (Reproduced by permission of the British Standards Institution, London).[1.2]

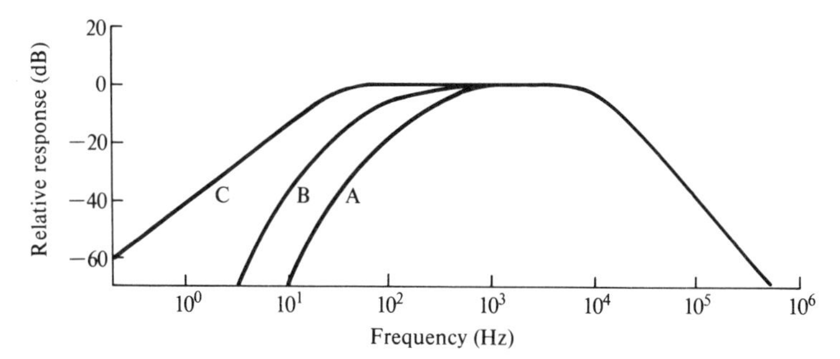

Fig. 1.4. A-, B-, and C-weighting curves.

where the sum is taken over all the m important components of the noise and suffix A represents the A-weighted value, i.e. the value adjusted by the A-weighting. The A-weighting adjustments at different frequencies are given in Table 1.2. Example 1.4 shows how to combine narrow-band sound-pressure levels to give the sound level in A-weighting. Figure 1.4 shows the A-weighting curve together with the B- and C-weighting curves. (The B- and C-weighting curves were originally used for louder sounds, the variations and sensitivity of average hearing over the audible frequency range being less for louder sounds. The A-weighting network is used much more than are the others.)

Example 1.4

The 1 per cent narrow-band sound-pressure levels of all predominant components of a noise are given in the following table. Estimate the A-weighted sound level.

Frequency (Hz)	1 per cent narrow-band sound pressure level, L_p (dB)	A-weighting adjustments (dB)	A-weighted sound pressure level, $L_{p,A}$ (dB)
100	96	−19.1	76.9
400	80	−4.8	75.2
800	78	−0.8	77.2
1000	80	0	80
1600	78	+1.0	79
2500	90	+1.3	91.3

The A-weighting adjustments can be found from Table 1.2. The A-weighted sound level is

$$L_A = 10 \log_{10}(10^{7.69} + 10^{7.52} + 10^{7.72} + 10^{8.0} + 10^{7.9} + 10^{9.13})$$
$$= 92.2 \text{ dB(A)}.$$

TABLE 1.2. Frequency weighting characteristics[1.5]†

Nominal frequency* (Hz)	Exact frequency* (Hz)	A-weighting (dB)	B-weighting (dB)	C-weighting (dB)
10	10.00	−70.4	−38.2	−14.3
12.5	12.59	−63.4	−33.2	−11.2
16	15.85	−56.7	−28.5	−8.5
20	19.95	−50.5	−24.2	−6.2
25	25.12	−44.7	−20.4	−4.4
31.5	31.62	−39.4	−17.1	−3.0
40	39.81	−34.6	−14.2	−2.0
50	50.12	−30.2	−11.6	−1.3
63	63.10	−26.2	−9.3	−0.8
80	79.43	−22.5	−7.4	−0.5
100	100.0	−19.1	−5.6	−0.3
125	125.9	−16.1	−4.2	−0.2
160	158.5	−13.4	−3.0	−0.1
200	199.5	−10.9	−2.0	−0.0
250	251.2	−8.6	−1.3	−0.0
315	316.2	−6.6	−0.8	−0.0
400	398.1	−4.8	−0.5	−0.0
500	501.2	−3.2	−0.3	−0.0
630	631.0	−1.9	−0.1	−0.0
800	794.3	−0.8	−0.0	−0.0
1 000	1 000	0	0	0
1 250	1 259	+0.6	−0.0	−0.0
1 600	1 585	+1.0	−0.0	−0.1
2 000	1 995	+1.2	−0.1	−0.2
2 500	2 512	+1.3	−0.2	−0.3
3 150	3 162	+1.2	−0.4	−0.5
4 000	3 981	+1.0	−0.7	−0.8
5 000	5 012	+0.5	−1.2	−1.3
6 300	6 310	−0.1	−1.9	−2.0
8 000	7 943	−1.1	−2.9	−3.0
10 000	10 000	−2.5	−4.3	−4.4
12 500	12 590	−4.3	−6.1	−6.2
16 000	15 850	−6.6	−8.4	−8.5
20 000	19 950	−9.3	−11.1	−11.2

* Nominal frequencies are as specified in ISO Standard 266. Exact frequencies are given above to four significant figures and are equal to $1000 \cdot 10^{n/10}$, where n is a positive or negative integer.

† (Reproduced by permission of the International Electrotechnical Commission, which retains the copyright).

The above example shows that the contribution of the peak component at 100 Hz to the A-weighted sound level is very little as there is a huge A-weighting attenuation of 19.1 dB at this frequency.

If noise measurements are made with octave- or one-third octave-band filters, the octave- or one-third octave-band sound-pressure levels can be combined in a way similar to that of the above example to obtain the A-weighted sound level.

1.8. A-weighted sound-power level

The noise limit for certain machinery specifications is sometimes specified in terms of A-weighted sound-power level, which is the combined total sound-power level with A-weighting adjustments of all individual sound-power levels at all the frequencies concerned. The A-weighted sound-power level can be expressed as

$$L_{\mathrm{W,A}} = 10 \log_{10} \sum_{i=1}^{m} 10^{0.1L_{\mathrm{W},i,\mathrm{A}}} \qquad (1.19)$$

where the sum is taken over all the m important components of the noise and the suffix A indicates the A-weighted value. Example 1.5 shows how to combine octave-band sound-power levels to give the A-weighted sound-power level.

Example 1.5

The octave-band sound-power levels for a machine are given in the following table. Calculate the A-weighted sound-power level.

Octave-band centre frequency (Hz)	Octave-band sound-power level, L_{W} (dB)	A-weighting adjustments (dB)	A-weighted sound-power level, $L_{\mathrm{W,A}}$ (dB)
63	30.5	−26.2	4.3
125	25.7	−16.1	9.6
250	23.8	−8.6	15.2
500	27.0	−3.2	23.8
1000	23.0	0	23.0
2000	29.4	+1.2	30.6
4000	25.4	+1.0	26.4
8000	23.9	−1.1	22.8

The A-weighted sound-power level is

$$\begin{aligned} L_{\mathrm{W,A}} &= 10 \log_{10}(10^{0.43} + 10^{0.96} + 10^{1.52} + 10^{2.38} + 10^{2.3} \\ &\quad + 10^{3.06} + 10^{2.64} + 10^{2.28}) \\ &= 33.5 \text{ dB(A)}. \end{aligned}$$

References

[1.1] Kinsler, L. E. and Frey, A. R. (1962). *Fundamentals of acoustics*, 2nd edn. John Wiley, New York.
[1.2] ISO (1961). *Normal equal loudness contours for pure tones and threshold of hearing under free-field listening conditions*, ISO R226. ISO, Geneva.
[1.3] ISO (1966). *Octave, half-octave and third-octave band filters intended for the analysis of sounds and vibrations*, ISO 225. ISO, Geneva.
[1.4] Fletcher, H. and Munson, W. A. (1933). Loudness, its definition, measurement and calculation. *J. acoust. Soc. Am.* **5,** 82.
[1.5] IEC (1979). *Sound level meters*, IEC 651. IEC, Geneva.

2. Noise and man

Noise may cause annoyance and damage to hearing and may affect speech communication. Furthermore, it may affect task performance and cause changes in the normal functions of the human organism. Our knowledge of the effects on task performance and organ function is very little. The following paragraphs will briefly discuss loudness, noise-induced hearing damage, and some common noise scales.

2.1. Loudness level and loudness

The loudness of a sound can be expressed either as *loudness* in *sones* or *loudness level* in *phons*. Loudness is measured on a noise scale designed to give scale values directly proportional to the loudness of a sound and its unit is *sone*. One sone is defined as the loudness experienced by a typical listener when listening to a pure tone of 1000 Hz having a sound-pressure level of 40 dB. A sound having a loudness of 2 sones would appear twice as loud as a sound of 1 sone when judged by a typical listener. (Experiments with juries of subjects show that 'twice as loud' does have about the same meaning for most subjects.)

Loudness level is another noise scale, having a unit of *phons*. The numerical value of the loudness level of a sound is defined as the numerical value of the sound-pressure level of a pure tone of 1000 Hz which is judged to be as loud as the sound. In other words, by definition, the loudness level of a 1000-Hz pure tone in phons is numerically equal to its sound-pressure level. The loudness level of other pure tones having a given sound-pressure level can be readily found from the equal loudness contours as shown in Fig. 1.3. For example, Fig. 1.3 shows that the loudness level of a pure tone of 100 Hz having a sound-pressure level of 29 dB is 10 phons since it is judged as loud as a 1000-Hz pure tone having a sound-pressure level of 10 dB.

It should be emphasized that the loudness level in phons does not give a value directly proportional to the loudness of a sound. For example, a sound having a loudness level of 40 phons is not twice as loud as another sound having a loudness level of 20 phons. Many tests with pure tones have shown that an increase in the loudness level of 10 phons corresponds approximately to a doubling in the loudness in sones. Thus the relationship between the loudness level (LL) in phons and the loudness (L) in sones is

$$L = 2^{(LL-40)/10}. \qquad (2.1)$$

The above expression together with Fig. 1.3 enable us to estimate the loudness of any pure tone if its sound-pressure level is given. However, Fig. 1.3 is based on pure tones and is therefore of no use for most machinery and industrial noises which usually contain many components at different frequencies. For evaluating the loudness of these complex noises, many different methods have been proposed. Two of the most widely used methods, Stevens' method and Zwicker's method, are recommended by the International Organization for Standardization.[2.1] The following paragraph will briefly present Stevens' method.

2.2. Stevens' method

Stevens' method for calculating the loudness of a complex noise is

(a) Measure the octave-band sound-pressure levels of the noise concerned;

(b) Find the loudness index value (LI) from Fig. 2.1 for each octave-band sound-pressure level;

(c) Determine the total loudness in sones by the empirical formula

$$L = (LI)_{max} + 0.3[\sum (LI) - (LI)_{max}] \tag{2.2}$$

where $(LI)_{max}$ is the highest loudness index value and $\sum (LI)$ is the sum of all the loudness indices.

If one-third octave-band sound-pressure levels are measured, the procedure is similar to the above, but the constant in eqn (2.2) should be 0.15 instead of 0.3.

Zwicker's method is given in ref. [2.1].

Example 2.1

The octave-band sound-pressure levels of a noise are given in the following table. Derive the loudness in sones.

Centre frequency of octave bands (Hz)	Octave-band sound pressure level (dB)	Loudness index (LI)
63	76	5
125	80	9.5
250	82	12.5
500	86	20.0
1000	96	50
2000	86	30
4000	77	20
8000	66	11.5
		$\sum (LI) = 158.5$

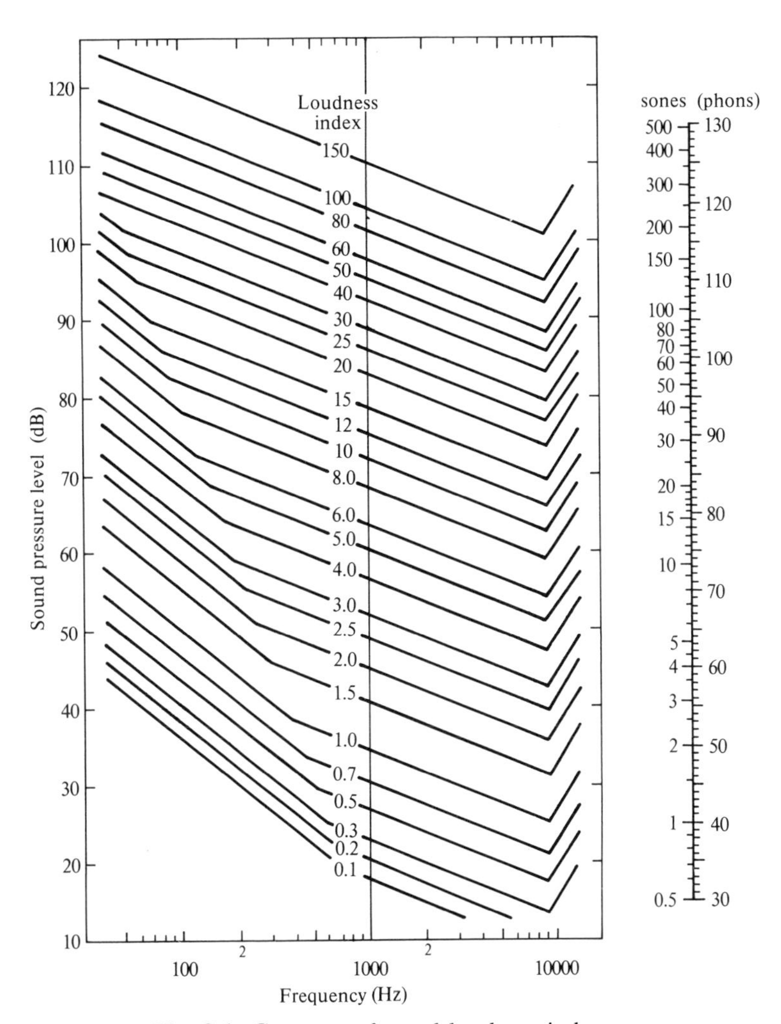

Fig. 2.1. Contours of equal loudness index.

The loudness indices can be found from Fig. 2.1 and are shown in the above table. The highest loudness index is $(LI)_{max} = 50$. The loudness of the noise is

$$L = 50 + 0.3(158.5 - 50)$$
$$= 82.6 \text{ sones.}$$

2.3. Noise-induced hearing damage

The sensitivity of the human ear deteriorates with noise exposure and age. For assessing noise-induced hearing damage the International Or-

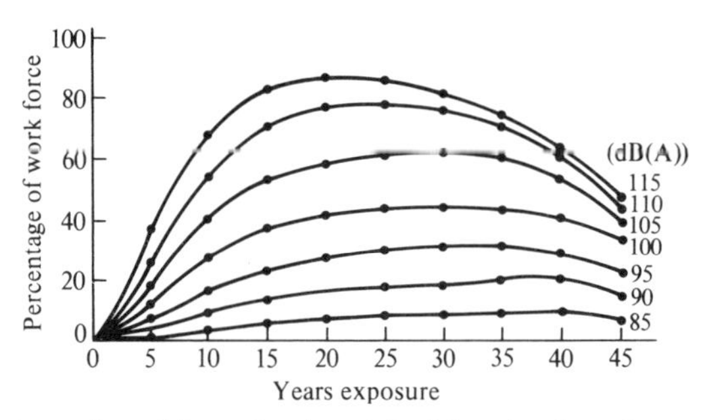

Fig. 2.2. Percentage of workforce having risk of hearing impairment for conversation due to noise as a function of equivalent continuous sound level in dB(A) during work and of exposure in years (Reproduced by permission of the British Standards Institution, London).[2.2]

ganization for Standardization has produced a document,[2.2] ISO 1999 *Assessment of occupational noise exposure for hearing conservation purposes.* The document gives the percentage of the work force expected to have impaired hearing for conversational speech as a function of noise exposure in terms of noise level in A-weighting and exposure duration (see Fig. 2.2). Here impaired hearing means exhibiting an increased threshold of hearing amounting to an average of 25 dB or more over the three main speech frequencies, 500 Hz, 1000 Hz, and 2000 Hz, due to noise exposure. Figure 2.2 shows that about 20 per cent of the work force is expected to have impaired hearing when exposed to a noise level of 90 dB(A) for 40 working years (based on a normal working week of 40 hours). Figure 2.2 does not apply to impulsive noises or high-level short-duration noises.

Table 2.1 shows the permissible noise exposures under the Walsh–

TABLE 2.1. Permissible noise exposures under Walsh–Healey Act

Duration per day (hours)	Sound level (dB(A)) slow response	
8	90	If daily noise exposure occurs in two or more periods at different levels, consider combined effect rather than individual effect of each. Sum up the fraction $C_1/T_1+C_2/T_2$, etc. where each C value is the total exposure time at a specified noise level, and each T value is the total time *permitted* at that level. If sum of fractions exceeds unity, then the mixed exposure exceeds the limit value.
6	92	
4	95	
3	97	
2	100	
$1\frac{1}{2}$	102	
1	105	
$1\frac{1}{2}$	110	
$\frac{1}{4}$ or less	115	

Healey Act of the USA. The Act permits a maximum daily 8-hour exposure limit of 90 dB(A) and an increase of 5 dB(A) for each halving of the exposure time. However, Burns and Robinson[2.3] suggested an increase of 3 dB(A) for each halving of the exposure time based on the equal energy concept. As regards impulse noise or impact noise, there is no international agreement on safe levels. Detailed discussion of the effects of noise on hearing is beyond the scope of this book and readers interested in the subject should refer to refs. [2.3] and [2.4].

In addition to the sound level in A-weighting, loudness level in phons, and loudness in sones, there are many other scales and units to describe a noise. The following paragraphs will introduce some of the more common ones.

2.4. Perceived noise level

The *perceived noise level* was developed by Kryter[2.5] and his colleagues for assessing aircraft noise. The procedure for calculating the perceived noise level is

(a) Measure the one-third octave-band or octave-band sound-pressure levels of the noise. If the noise is not a steady one, e.g. aircraft fly-over noise, find the maximum sound-pressure levels attained in each one-third octave band or octave band during the event, regardless of whether these maximum levels in different bands occur simultaneously.

(b) Use Fig. 2.3 and the maximum sound-pressure level in each band to find the noisiness in noys for each band.

(c) Determine the total *perceived noisiness PN* in noys using the equation

$$PN = N_{max} + F\left(\sum N - N_{max}\right) \tag{2.3}$$

where N_{max} is the highest noisiness value, $\sum N$ is the sum of the noisiness values in all bands, and F is 0.3 for octave bands or 0.15 for one-third-octave bands.

(d) The perceived noise level, L_{PN}, with a unit of PN dB, is given by

$$L_{PN} = 40 + 10 \log_2(PN). \tag{2.4}$$

If predominant pure tones and other pronounced irregularities are present in the noise spectrum, a correction should be made and the result is called the *tone-corrected perceived noise level*, PNLT. Based on the PNLT values and taking into account the signal duration of the event, one can find the *effective perceived noise level*, EPN(dB), as defined by ISO 3891.[2.6] The corrections and procedures to obtain the tone-corrected perceived noise level and the effective perceived noise level are rather complicated and the details can be found in ref. [2.6].

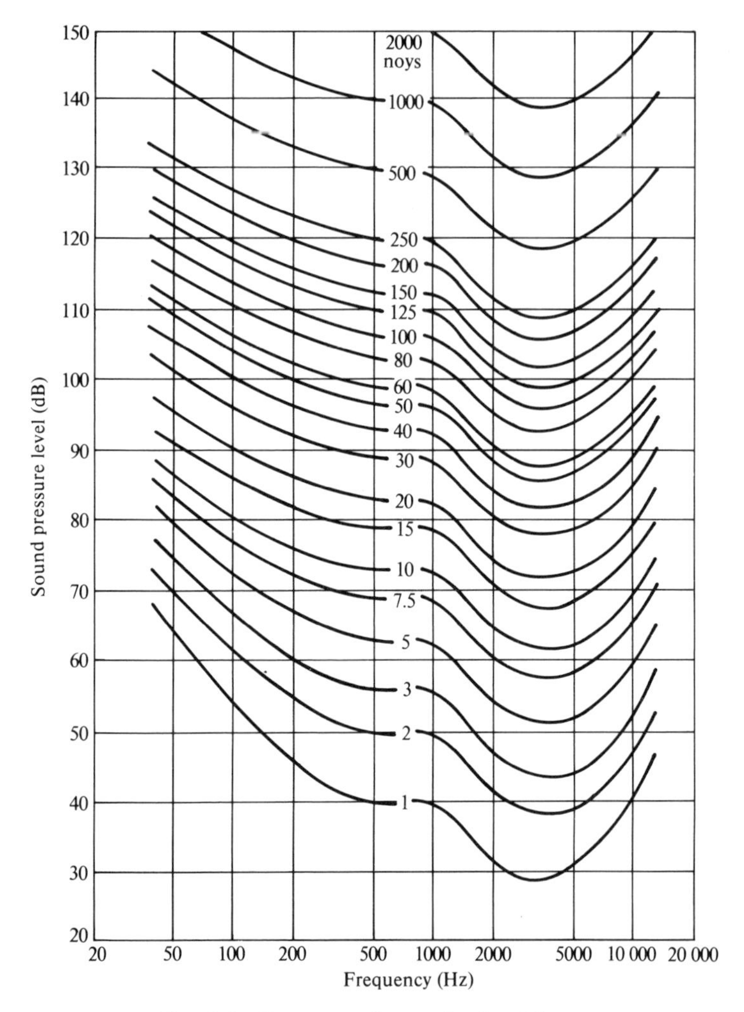

Fig. 2.3. Contours of perceived noisiness.

2.5. Traffic noise index

A noise scale called the *traffic noise index*[2.7,2.8] has been proposed to evaluate the subjective response to time-varying road traffic noise. The base measure is the A-weighted sound level recorded outdoors continuously or sampled at numerous discrete intervals over a 24-hour period. From the recorded data of sound level (A) versus time (see Fig. 2.4) one can determine the L_{10} level, the sound level (A) which was exceeded for 10 per cent of the time, and the L_{90} level, the sound level (A) exceeded for 90 per cent of the time.

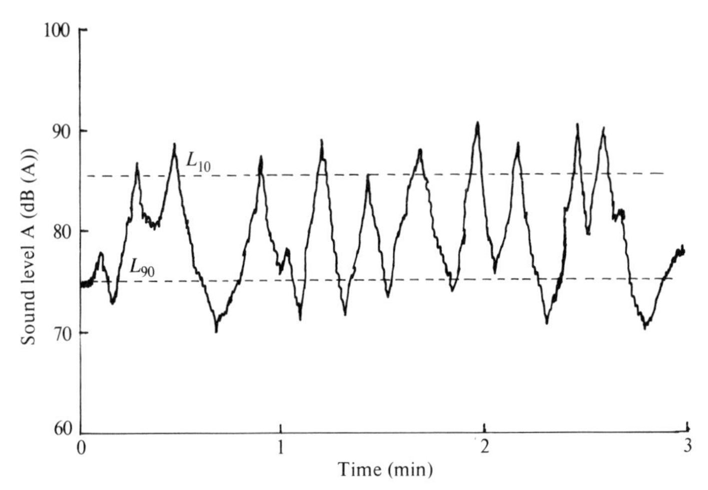

Fig. 2.4. Example of time-varying traffic noise. L_{10} = sound level in dB(A) exceeded for 10 per cent of the time; L_{90} = sound level in dB(A) exceeded for 90 per cent of the time.

The traffic noise index *TNI* is defined as

$$TNI = 4(L_{10} - L_{90}) + L_{90} - 30. \tag{2.5}$$

The first term describes the variability of the noise, the second term represents the average 'background' noise, and the third term is introduced to give more convenient numbers.

2.6. Equivalent continuous sound level

The *equivalent continuous A-weighted sound level* for a given time period from t_1 to t_2 is defined as

$$L_{\text{Aeq}} = 10 \log_{10}\left(\frac{1}{t_2 - t_1}\int_{t_1}^{t_2} 10^{L_{\text{A}}(t)/10}\, dt\right) \tag{2.6}$$

where $L_{\text{A}}(t)$ is the instantaneous sound level (A). The value in the bracket of eqn (2.6) gives the mean value of $p_{\text{A}}^2/p_{\text{ref}}^2$ over the period concerned. (p_{A} is the A-weighted r.m.s. sound pressure and p_{ref} is the reference sound pressure of 2×10^{-5} N m^{-2}.) Since the sound intensity is approximately proportional to the square of sound pressure, the *equivalent continuous sound level* is the constant sound level which would expose the ear to the same amount of A-weighted sound energy as does the actual time-varying sound over the same period. Therefore L_{Aeq} is also called the energy equivalent sound level. If the time interval is 24 hours or 8 hours, then L_{Aeq} is designated as $L_{\text{Aeq(24)}}$ or $L_{\text{Aeq(8)}}$, respectively. von

Gierke[2.9] considered L_{Aeq} to be the best descriptor for the magnitude of the environmental noise. Many noise-measuring meters giving L_{Aeq} readings based on adjustable time intervals are now available commercially.

2.7. Noise pollution level

There is some evidence that annoyance depends on the variability of a noise and on the equivalent continuous sound level. The *noise pollution level*, L_{NP}, is based on these two and is expressed as[2.10]

$$L_{NP} = L_{Aeq} + K\sigma$$

where L_{Aeq} is the equivalent continuous sound level in A-weighting, σ is the standard deviation of the A-weighted sound levels at discrete time intervals over a given period, and K is a constant tentatively set equal to 2.56 as this value leads to the best fit with currently available studies of subjective response to noise.

For many community noises, one can use the alternative expression for the noise pollution level

$$L_{NP} = L_{Aeq} + (L_{10} - L_{90}) \tag{2.8}$$

where L_{10} and L_{90} are the A-weighted sound levels exceeded for 10 and 90 per cent of the time, respectively.

The above paragraphs have introduced a few noise scales of major interest and there are many more different noise scales and units in the literature. It is impossible to say that one particular scale is the best and a choice is usually made in terms of the particular application. However, the A-weighted sound level is the best known and most widely applicable noise scale. For a steady noise without predominant pure tones, the A-weighted sound level is a convenient and satisfactory scale for rating noise annoyance and hearing loss. For noises having predominant pure tones and for impulsive and intermittent noises (such as typewriter noise), the A-weighted sound level is not a satisfactory scale for assessing the subjective response of human beings. For machinery noise measurement and control, the A-weighted sound level referred to a reference distance from the machine surface and the A-weighted sound-power level are the two values most frequently used in practice.

References

[2.1] ISO (1975). *Method for calculating loudness level*, ISO 532. ISO, Geneva.

[2.2] ISO (1975). *Assessment of occupational noise exposure for hearing conservation purpose*, ISO 1999. ISO, Geneva.

[2.3] Burns, W. and Robinson, D. W. (1970). *Hearing and noise in industry*. HMSO, London.

[2.4] Davis, H. and Silverman, S. R. (1970). *Hearing and deafness.* Holt, Rinehart and Winston, New York.

[2.5] Kryter, K. D. (1968). Concepts of perceived noisiness, their implementation and application. *J. acoust. Soc. Am.* **43,** 344–461.

[2.6] ISO (1978). *Procedure for describing aircraft noise heard on the ground,* ISO 3891. ISO, Geneva.

[2.7] Langdon, F. J. and Scholes, W. E. (1968). *The traffic noise index: a method of controlling noise nuisance.* CP 38/68, Building Research Station, Watford.

[2.8] Griffiths, I. D. and Langdon, F. J. (1968). Subjective response to road traffic noise. *J. Sound Vibration* **8** (1), 16–32.

[2.9] von Gierke, H. E. (1974). Noise—how much is too much? In *The 8th International Congress on Acoustics, London, Invited Lectures,* pp. 149–76. Goldcrest Press, Trowbridge.

[2.10] Robinson, D. W. (1969). *The concept of noise pollution level.* NPL Aero-Report Ac 38, National Physical Lab., Teddington.

3. Sound power measurement

3.1. Noise field around a machine

The noise field around a machine varies in general with direction, time, the acoustic environment, and the operating and mounting conditions. For example, the variation of sound-pressure level at 100 Hz in the horizontal plane through the shaft of an electric machine resiliently supported at the centre of an anechoic chamber is shown in Figs. 3.1 and 3.2.[3.1] Figure 3.1 shows the polar diagrams at various distances from the machine centre and Fig. 3.2 gives the variation of sound-pressure level with distance at various angles. Figure 3.2 also shows a dotted 'inverse-square law' line, representing the ideal variation of sound-pressure level in free field if the noise power were radiating evenly in all directions. One sees that the shapes of the polar diagrams are not the same and the variation of sound-pressure level with distance along a particular direction deviates irregularly from the 'inverse-square law' line. These complicated noise-field pictures could be explained by the following facts: (1) the machine surface was of an irregular shape; (2) the machine noise had various origins—aerodynamic, electromagnetic, and mechanical. Different origins would emit noise in different directions and could cause surface vibrations at various modes of vibration to produce complex noise patterns.

The noise field of a machine is even more complicated if there are one or more sound reflecting surfaces near the machine. The reflected sound waves from these surfaces would interact with the direct sound waves. If a direct sound wave and a reflected sound wave are in phase at certain points, then the sound pressure at these points is strengthened. On the other hand, if the direct and reflected waves are in phase opposition, the sound pressure is weakened. The effect of a reflecting surface on a sound field depends mainly on the wavelength of the sound and the distance of the noise source from the reflecting surface; this will be discussed in Section 4.3.

Although a large number of machines produce steady noise under specified operating and mounting conditions, some machines emit non-steady or pulsating noise. For example, a two-pole single-phase electric motor usually emits a non-steady noise[3.3] (see Fig. 3.3), which is of electromagnetic origin. Pneumatic tools, office machines, and data-processing equipment also emit non-steady impulsive noise. Figure 3.4 gives an example of typewriter impulsive noise.

The noise emitted from a machine also varies with its operating

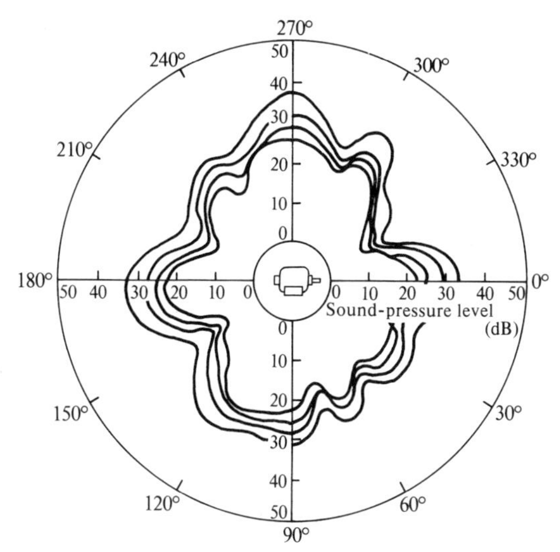

Fig. 3.1. Variation of sound-pressure level along radial lines for a pure-tone component from an electric machine.[3.1] (a) $r = 0.3$ m (r = distance from machine centre); (b) $r = 0.46$ m; (c) $r = 0.61$ m; (d) $r = 0.76$ m.

conditions. Figure 3.5 shows the noise-frequency spectra of a car-alternator on no load and at full load, respectively, when running at the same speed.

The mounting conditions of a machine will affect the natural frequencies and the structure-borne noise emission. This is of considerable importance if a machine is connected through its mountings to a thin-plate type structure. Part of the machine vibrations will be transmitted to

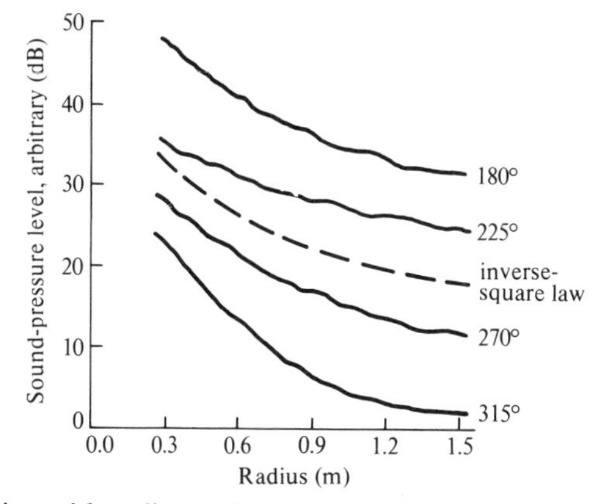

Fig. 3.2. Variation with radius and angular direction of sound-pressure level for a pure-tone component from an electric machine.[3.1]

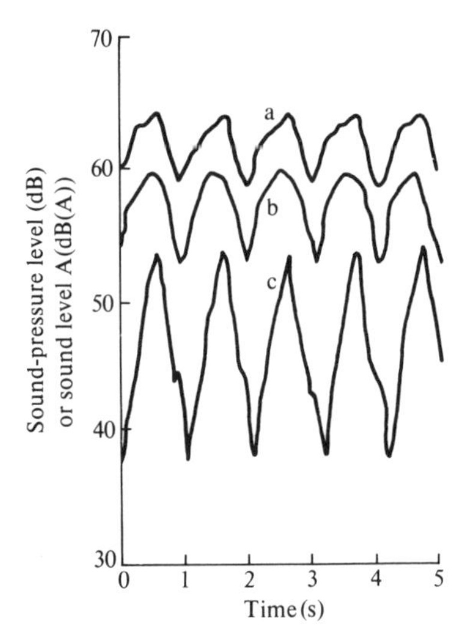

Fig. 3.3. Examples of measurement of pulsating noise emitted from a two-pole single-phase machine in an anechoic chamber.[3.2] (a) In A-weighting, speed = 2970 rev min^{-1}; (b) in one-third octave band, centred at 1250 Hz, speed = 2970 rev min^{-1}; (c) 1239 Hz, passband 1 per cent of centre frequency, speed = 2970 rev min^{-1}.

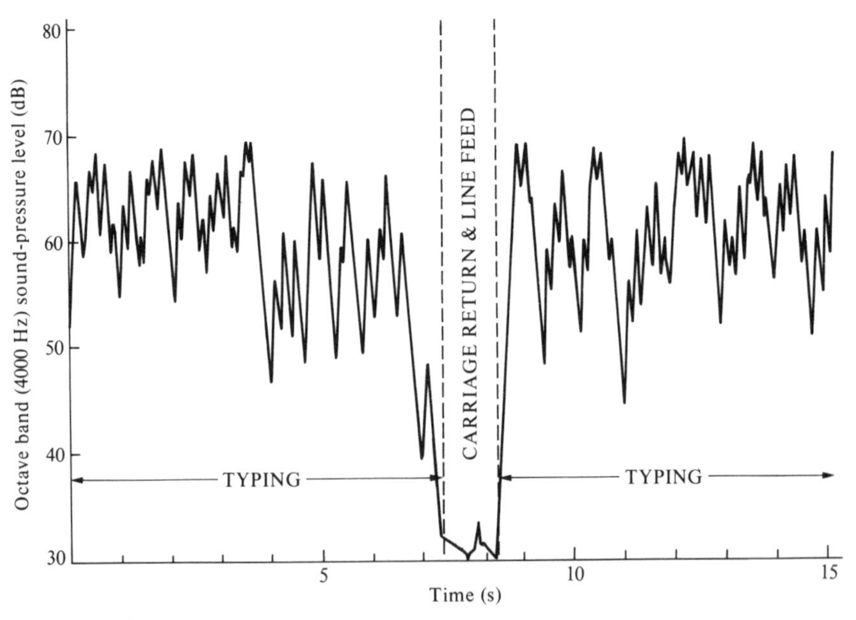

Fig. 3.4. Example of electric typewriter noise. Variation of sound-pressure level (octave-band centred at 4000 Hz) with time at operator's position.

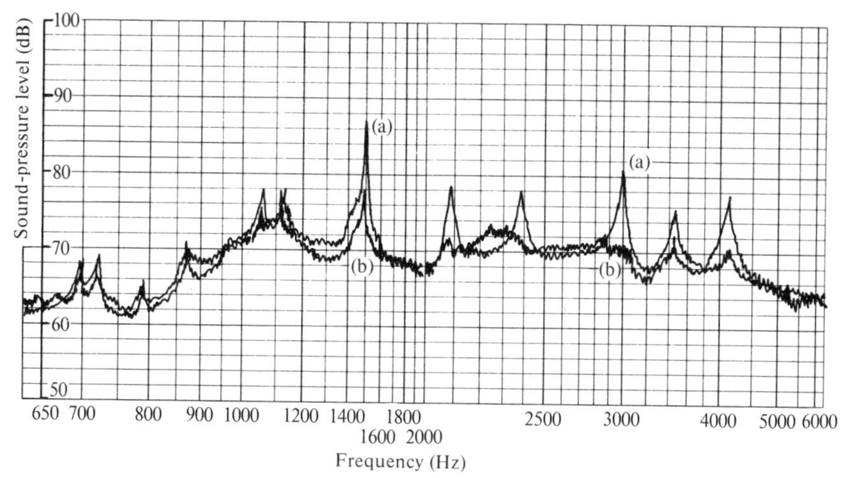

Fig. 3.5. Frequency analysis of noise from a car alternator at the same point in noise field (speed = 7000 rev min^{-1}, 1 per cent narrowband). (a) With full-load current; (b) with no-load current.

the thin-plate structure, which will in turn emit 'additional' noise, especially when the machine vibration frequencies are at or near some of the natural frequencies of the structure.

The above discussions show clearly that, because of the nature of the noise field around a machine, a single-point sound-pressure level measurement is inadequate to describe the noise emitted from a machine, and quoting a noise-level result without specifying the measuring, operating, and mounting conditions would be almost meaningless.

3.2. Sound-power measurement in free field

Section 3.1 shows that the sound-pressure level values around a machine vary from point to point irregularly. However, if sound-pressure level measurements are made at a number of points on a surface enclosing a machine, it is possible to find a unique value for the sound power in each frequency band emitted from the machine under specified operating and mounting conditions.

3.2.1. *Spherical surface measurement*

When accurate noise power from a machine is required, the machine should be placed at the centre of an anechoic chamber or suspended in a large open area and multiple-point sound-pressure level measurements made over a spherical surface enclosing the machine. The following assumptions are necessary.

(1) The errors introduced by the measuring equipment and personnel are negligible.

(2) The number of measuring points n is very large.

(3) The measurements are in the far field, i.e. the sound pressure and the particle velocity are approximately in phase.

(4) The spherical measuring surface having radius r is normal to the direction of sound propagation.

The sound-power level is then, from eqn (1.9),

$$L_W = 10 \log_{10}\left(\frac{(p_i^2)_{av}}{p_{ref}^2}\right)_{sph} + 10 \log_{10} 4\pi r^2 = \bar{L}_{p,sph} + 10 \log_{10} 4\pi r^2 \quad (3.1)$$

TABLE 3.1. Recommended array of microphone positions in a free field[3.16]*

No.	$\frac{x}{r}$	$\frac{y}{r}$	$\frac{z}{r}$
1	−0.99	0	0.15
2	0.50	−0.86	0.15
3	0.50	0.86	0.15
4	−0.45	0.77	0.45
5	−0.45	−0.77	0.45
6	0.89	0	0.45
7	0.33	0.57	0.75
8	−0.66	0	0.75
9	0.33	−0.57	0.75
10	0	0	1.0
11	0.99	0	−0.15
12	−0.50	0.86	−0.15
13	−0.50	−0.86	−0.15
14	0.45	−0.77	−0.45
15	0.45	0.77	−0.45
16	−0.89	0	−0.45
17	−0.33	−0.57	−0.75
18	0.66	0	−0.75
19	−0.33	0.57	−0.75
20	0	0	−1.0

The locations of 20 points associated with equal areas on the surface of a sphere of radius r are shown in this table which gives the Cartesian coordinates (x, y, z) with origin at the centre of the source. The z-axis is chosen perpendicularly upward from a horizontal plane $(z = 0)$.

* (Reproduced by permission of the British Standards Institution, London.)

where

$$\bar{L}_{p,sph} = 10 \log_{10}\left(\frac{1}{n}\sum_{i=1}^{n} 10^{L_{p,i}/10}\right)_{sph} = 10 \log_{10}\left(\frac{(p_i^2)_{av}}{p_{ref}^2}\right)_{sph}. \quad (3.2)$$

$\bar{L}_{p,sph}$ is called the *spherical-equivalent sound-pressure level.*

In practice the number of measuring points is rarely greater than 20. The possible error in the sound-power level due to an insufficient number of measuring points and other reasons will be discussed in Chapter 4. Table 3.1 gives the coordinates of uniformly disposed measuring points on a spherical surface. According to ISO 3745,[3.16] the radius of the measuring spherical surface shall be equal to or greater than twice the major source dimension, but not less than 1 m.

An example of the variation with distance of the spherical-equivalent sound-pressure level produced by an electric machine in an anechoic chamber is given in Fig. 3.6. Figure 3.6 shows that, except for the 100-Hz component where the noise measurements were made in the near field, the variation of the spherical-equivalent sound-pressure level, hence the sound-power level, conforms to the inverse-square law shown by the straight lines even though the variations of sound-pressure level along radial lines do not, as shown in Fig. 3.2.

Since most machines do not radiate noise uniformly in all directions, the sound-pressure level measured at a radius r and at a particular

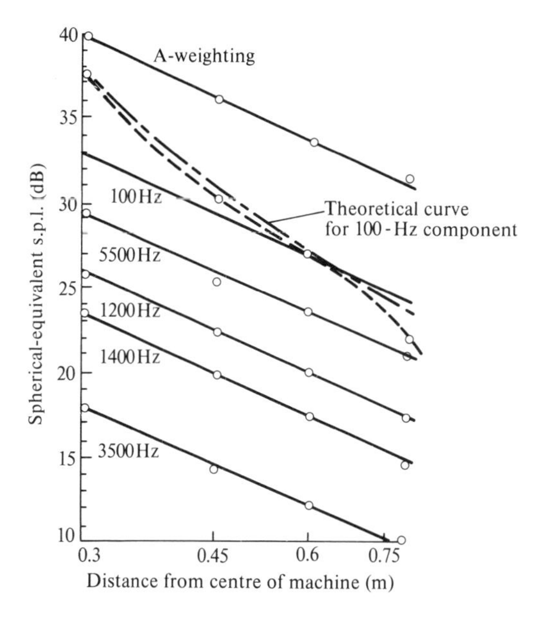

Fig. 3.6. Variation of spherical-equivalent sound-pressure level with distance.[3.1]

location i is usually different from the spherical-equivalent sound-pressure level on the spherical surface of radius r. The amount by which the sound-pressure level in a particular location i exceeds the spherical-equivalent sound-pressure level is called the *directivity index*, i.e.

$$DI_i = L_{p,i} - \bar{L}_{p,sph}. \tag{3.3}$$

Example 3.1

The 1250-Hz one-third octave-band sound-pressure levels obtained in an anechoic chamber on a 1.2-m radius spherical surface enclosing a machine, using a 12-point array, are 76, 78, 79, 82, 83, 86, 88, 88, 89, 89, 90, and 92 dB. Calculate (1) the spherical equivalent sound-pressure level; (2) the sound-power level; and (3) the directivity index at the location having a sound-pressure level of 92 dB.

Solution.

Sound-pressure level, L_p (dB)	$10^{L_p/10} = \frac{p^2}{p^2_{ref}}$
76	3.98×10^7
78	6.30×10^7
79	7.94×10^7
82	1.58×10^8
83	1.99×10^8
86	3.98×10^8
88	6.30×10^8
88	6.30×10^8
89	7.94×10^8
89	7.94×10^8
90	1.00×10^9
92	1.58×10^9

$$\sum_{i=1}^{12} 10^{L_{p,i}/10} = 6.37 \times 10^9$$

$$\frac{(p_i^2)_{av}}{p^2_{ref}} = \frac{1}{12}\left(\sum_{i=1}^{12} 10^{L_{p,i}/10}\right) = \tfrac{1}{12}(6.37 \times 10^9) = 5.31 \times 10^8.$$

From eqn (3.2), the spherical-equivalent sound-pressure level $\bar{L}_{p,sph}$ in the 1250-Hz one-third octave-band is

$$\bar{L}_{p,sph} = 10 \log_{10}(\tfrac{1}{12}(6.37 \times 10^9)) = 87.2 \text{ dB}.$$

From eqn (3.1), the sound-power level in the band is

$$L_W = 87.2 + 10 \log_{10}[4\pi(1.2)^2] = 99.8 \text{ dB}.$$

From eqn (3.3), the directivity index at the point having a sound-pressure level of 92 dB is

$$DI = 92 - 87.2 = 4.8 \text{ dB}.$$

3.2.2. *Hemispherical surface measurement*

Sound-power measurements are sometimes made with the machine mounted on the ground in an open area or in a semi-anechoic chamber, i.e. an anechoic chamber having a hard (reflecting) floor. Sound-pressure level measurements can be made over a hemispherical surface of radius r enclosing it. According to ISO 3745,[3.16] the radius of the hemispherical surface shall be equal to or greater than twice the major source dimension, or four times the average distance of the source from the reflecting floor, whichever is the larger, and not less than 1 m. Table 3.2 gives the coordinates of 10 key measurement points. Figure 3.7 illustrates the relative locations of microphone positions.

Based on eqn (1.9), assuming no sound power transmitted through or absorbed by the ground and neglecting the effects of ground reflection, the sound-power level can be expressed as

$$L_W = \bar{L}_p + 10 \log_{10} 2\pi r^2 \tag{3.4}$$

where $\bar{L}_p$ is defined by eqn (1.10).

TABLE 3.2. Coordinates of measuring points on a hemispherical surface[3.16]†

(a) Coordinates of key measurement points

No.	x/r	y/r	z/r
1	−0.99	0	0.15
2	0.50	−0.86	0.15
3	0.50	0.86	0.15
4	−0.45	0.77	0.45
5	−0.45	−0.77	0.45
6	0.89	0	0.45
7	0.33	0.57	0.75
8	−0.66	0	0.75
9	0.33	−0.57	0.75
10	0	0	1.0

(b) Recommended microphone positions when the source emits predominant pure tones*

No.	x/r	y/r	z/r
1	0.16	−0.96	0.22
2	0.78	−0.60	0.20
3	0.78	0.55	0.31
4	0.16	0.90	0.41
5	−0.83	0.32	0.45
6	0.83	−0.40	0.38
7	−0.26	−0.65	0.71
8	0.74	−0.07	0.67
9	−0.26	0.50	0.83
10	0.10	−0.10	0.99

* If the source emits predominant pure tones, strong interference effects may occur if several microphone positions are placed at the same height above the reflecting plane. In such cases the use of a microphone array with the coordinates given in (b) is recommended.

† (Reproduced by permission of the British Standards Institution, London.)

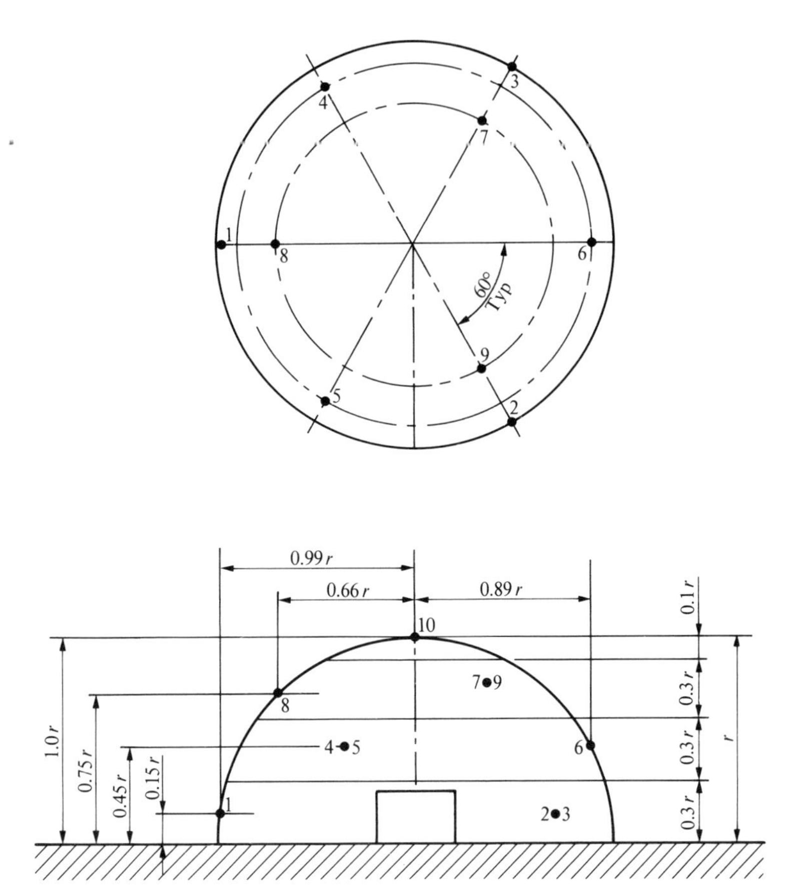

Fig. 3.7. Location of microphones for a 10-point hemispherical measurement (Reproduced by permission of the British Standards Institution, London).[3.16]

3.2.3. *Conformal and other measurement surfaces*

For long and thin or very large machines it is difficult to take either spherical or hemispherical measurements.

ISO Recommendation 1680[3.4] and BEAMA Publication No. 225[3.5] suggest measurement of the sound-pressure level at a number of points which are at a constant distance from the machine surface (see Fig. 3.8(a)) along prescribed paths in both vertical and horizontal planes. The sound-power level is given by

$$L_W = \bar{L}_p + 10 \log_{10} 2\pi r_{eq}^2 \tag{3.5}$$

where the equivalent radius r_{eq} is

$$r_{eq} = \left(\frac{a(b+c)}{2}\right)^{1/2}$$

where a, b, and c are as shown in Fig. 3.8(a), (b).

The above r_{eq} expression is derived from the assumption that the

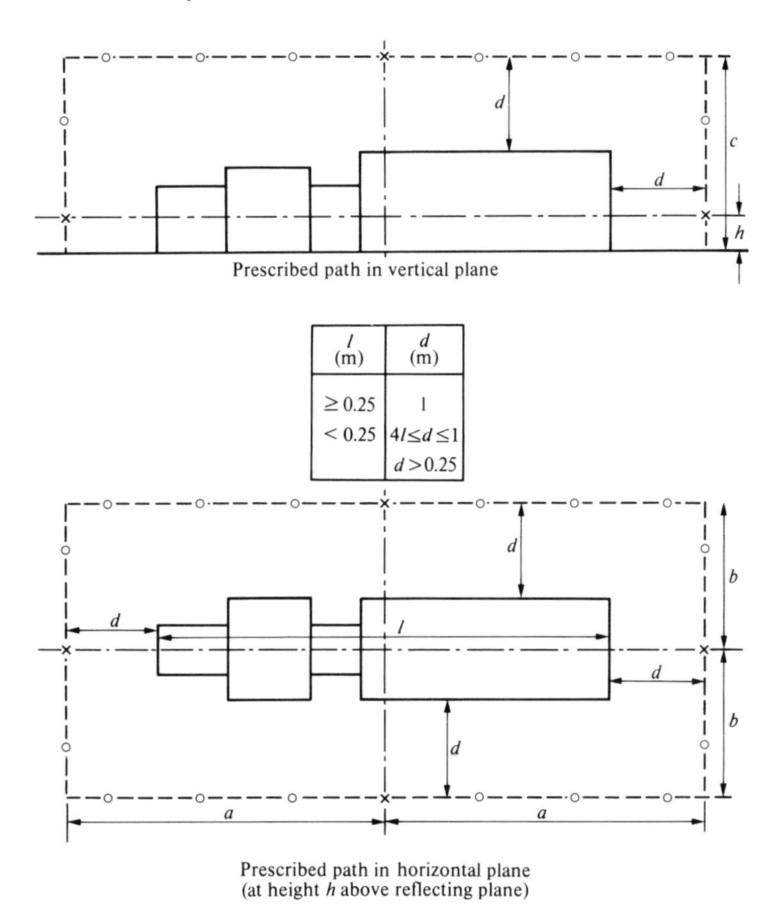

h = shaft height or 0.25 metre wichever is greater

○ - key measuring point

× - other measuring points marked off at intervals of 1 m from key points

Fig. 3.8. (a) Location of measuring points and prescribed paths for horizontal machines.[3,5]

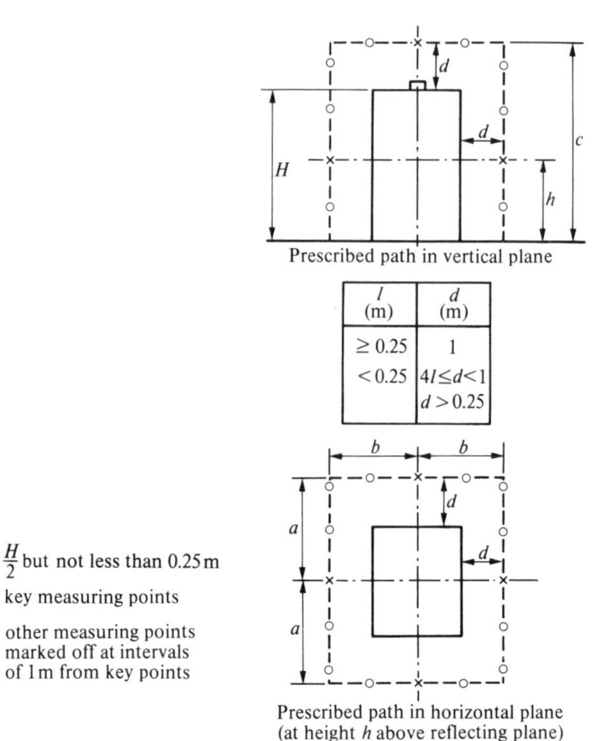

l (m)	d (m)
≥ 0.25	1
< 0.25	$4l \leq d < 1$ $d > 0.25$

Fig. 3.8. (b) Location of measuring points and prescribed paths for vertical machines.[3,5]

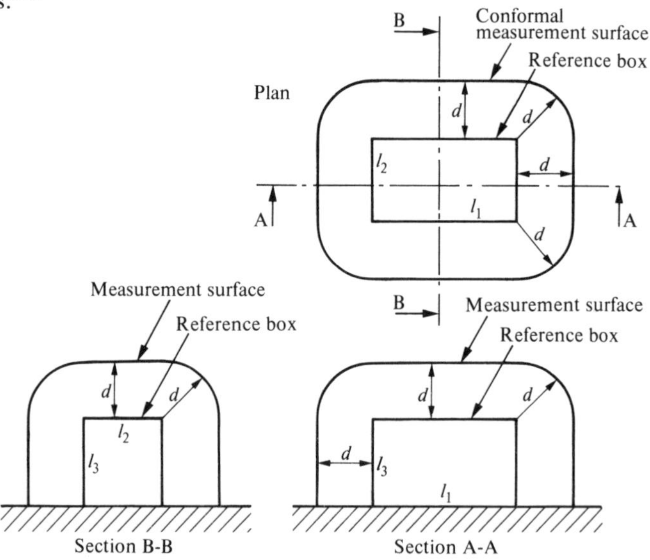

Fig. 3.8. (c) Reference box and conformal surface (reproduced by permission of the British Standards Institution, London).[3,18]

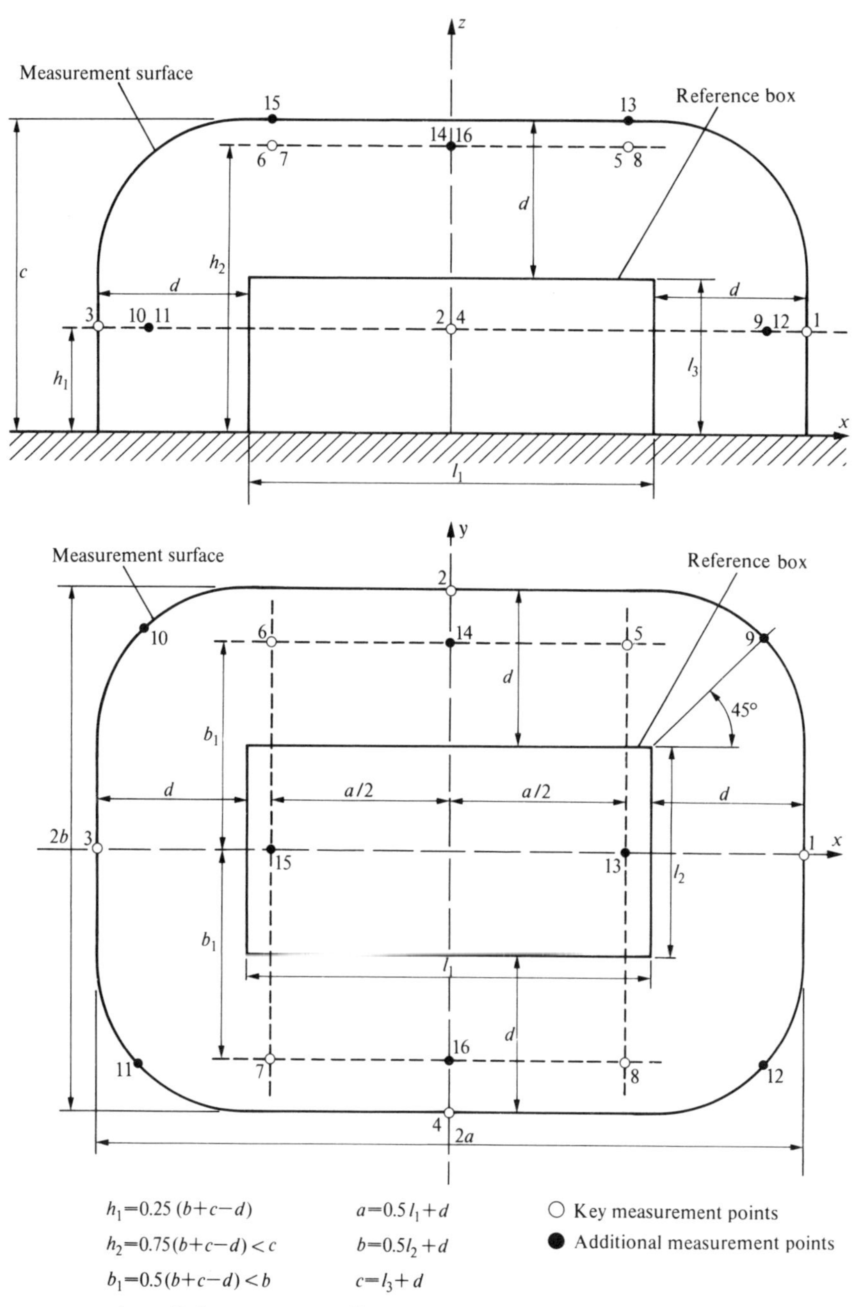

$h_1 = 0.25\,(b+c-d)$ $a = 0.5\,l_1 + d$ ○ Key measurement points

$h_2 = 0.75\,(b+c-d) < c$ $b = 0.5\,l_2 + d$ ● Additional measurement points

$b_1 = 0.5\,(b+c-d) < b$ $c = l_3 + d$

where d is the measurement distance, normally 1 m.

The minimum height of the microphone above the ground shall be 0.15 m.

Fig. 3.8. (d) Microphone positions on the conformal surface (reproduced by permission of the British Standards Institution, London).[3,18]

semi-circular ends of the measuring surface are negligible. However, a comparison between this method and the hemispherical method has shown that the area of the semi-circular ends is not negligible for machines having approximately square shapes.[3.1] If the semi-circular ends are to be included, the equivalent radius r_{eq} is

$$r_{eq} = \left[\frac{b+c}{2}\left(a + \frac{b+c}{4}\right)\right]^{1/2}.$$

According to ISO 3744,[3.18] it is necessary first to establish a reference box for the source. The reference box is the smallest possible rectangular box that just encloses the source and terminates on the reflecting plane. If the dimensions of the box l_1, l_2, and l_3 (see Fig. 3.8(c)) are less than 1 m, the hemispherical surface is preferred. If any dimension of the box exceeds 1 m and the reference box is not approximately cubical in shape, then the conformal (or parallelepiped) measurement surface is preferred.

The conformal surface is that surface which is defined as being located everywhere a distance d from the nearest point on the envelope of the reference box (see Fig. 3.8(c)). It is an enclosure formed by a rectangular parallelepiped with round corners, the corners being formed by portions of cylinders and spheres. The measurement distance d is the perpendicular distance to the reference box from the side of the conformal surface. The eight key microphone positions are shown in Fig. 3.8(d).

The sound-power level is given by

$$L_W = \bar{L}_p + 10 \log_{10} S_{con}, \tag{3.6}$$

where $\bar{L}_p$ is defined by eqn (1.10) and S_{con} is the area of the conformal surface which is given approximately by

$$S_{con} = 4(ab + bc + ca) \times \frac{a+b+c}{a+b+c+2d} \tag{3.7}$$

where $a = 0.5l_1 + d$, $b = 0.5l_2 + d$, $c = l_3 + d$, and d is the measurement distance, normally 1 m. The value of d shall preferably be one of 0.25, 0.5, 1, 2, 4, or 8 m. If any dimension of the reference box is larger than $2d$, or the spread of sound-pressure level values exceeds 8, eight additional microphone positions should be used (see Fig. 3.8(d)).

If the source radiates noise with a high directivity or if the noise is mainly from a small portion, e.g. the openings of an otherwise enclosed machine, additional measurement positions in the region of high noise radiation shall be used.

3.3. Sound-power measurement in semi-reverberant spaces

As anechoic chambers and reverberant rooms are either not available or too small for a bulky noise source, sound-pressure measurements have to be made sometimes in a semi-reverberant space, i.e. in an ordinary room, which is neither anechoic nor diffuse and in which there are often other objects in addition to the noise source under test. It is therefore necessary to include a correction for room reflection in sound-power determination. The following paragraphs will describe how to determine the machinery sound power in a semi-reverberant space.

3.3.1. *Reference sound-source substitution method*

A reference sound source of known sound-power levels in various frequency bands of interest can be used for calibrating the room reflection in a semi-reverberant space. The difference between the true power of the source and the measured power in a semi-reverberant space for a particular measuring surface is then taken as the room reflection correction for that measuring surface. If sound-pressure level measurements are made on a specified measuring surface for both the machine under test and the reference noise source placed at the same location, the sound-power level of the machine under test can be found from

$$L_W = L_{W,ref} + \bar{L}_p - \bar{L}_{p,ref} \tag{3.8}$$

where L_W = the sound-power level of the machine under test in a frequency band, $L_{W,ref}$ = the known sound-power level of the reference sound source in the same frequency band (determined by calibration in a free field), $\bar{L}_p$ = the mean level of sound pressure over a measuring surface for the machine under test (determined by eqn (1.10)), and $\bar{L}_{p,ref}$ = the mean level of sound pressure obtained over the same measuring surface in the semi-reverberant space for the reference sound source.

This method can provide satisfactory sound-power results which are accurate enough for most engineering applications and it is not necessary to find the room characteristic parameter for sound absorption in terms of the room constant or reverberation time. However, a free field should be used to calibrate the reference noise source immediately before using it and the condition for maximum accuracy is that the general shape of the sound spectrum of the reference sound source and the machine under test should be similar.[3.4] The size of the machine and that of the reference sound source should also be of the same order. These necessary conditions restrict considerably the use of this reference-source substitution method.

Example 3.2

Sound-pressure levels in the 2000-Hz octave band in a semi-reverberant space using an eight-point hemispherical array enclosing a machine are 80, 84, 86, 88, 89, 92, 94, and 95 dB. After replacing the machine by a reference sound source at the same location, sound-pressure levels at the same measuring points in the 2000-Hz band are 76, 78, 79, 82, 83, 86, 88, and 89 dB. Measurements on the reference sound source in an anechoic chamber result in a sound-power level of 83 dB in the 2000-Hz octave band. Derive the sound-power level of the machine in the octave band.

Solution. Based on eqn (1.10), the mean sound-pressure level on the hemispherical surface is

$$\bar{L}_p = 10 \log_{10}\left(\frac{1}{8}\sum_{i=1}^{8} 10^{0.1L_{p,i}}\right) = 10 \log_{10}(\tfrac{1}{8}(9.42\times10^9))$$

$$= 90.7 \text{ dB}.$$

The mean sound-pressure level for the reference sound source on the same surface is

$$\bar{L}_{p,ref} = 10 \log_{10}\left(\frac{1}{8}\sum_{i=1}^{8} 10^{0.1L_{p,i}}\right)_{ref} = 10 \log_{10}(\tfrac{1}{8}(2.36\times10^9))$$

$$= 84.7 \text{ dB}.$$

From eqn (3.8), the sound-power level of the machine in the 2000-Hz octave band is

$$L_W = L_{W,ref} + \bar{L}_p - \bar{L}_{p,ref} = 83 + 90.7 - 84.7 = 89 \text{ dB}$$

3.3.2. *'Close-field' measurement method*

The effect of room reflection on the sound-pressure levels in the vicinity of a small machine situated in a relatively large room would be expected to be small. Based on this, King[3.6] suggests that it is possible to determine the noise output of a small machine in an ordinary room by making sound-pressure level measurements at locations close to the machine surface. For a machine of about 0.31 m diameter, King found that at a distance of 0.31 m from the surface of the machine the effect of the reflected waves in a large semi-reverberant room of 10 m × 4 m × 5 m was negligible. Further studies[3.1] showed that the sound-power levels in one-third octave bands obtained from sound-pressure level measurements made in an ordinary laboratory at points 0.3 m from the surface of a small electric machine were approximately the same as the power levels obtained from sound-pressure level measurements in an anechoic chamber for frequencies higher than 315 Hz (see Fig. 3.9). However, Fig. 3.9

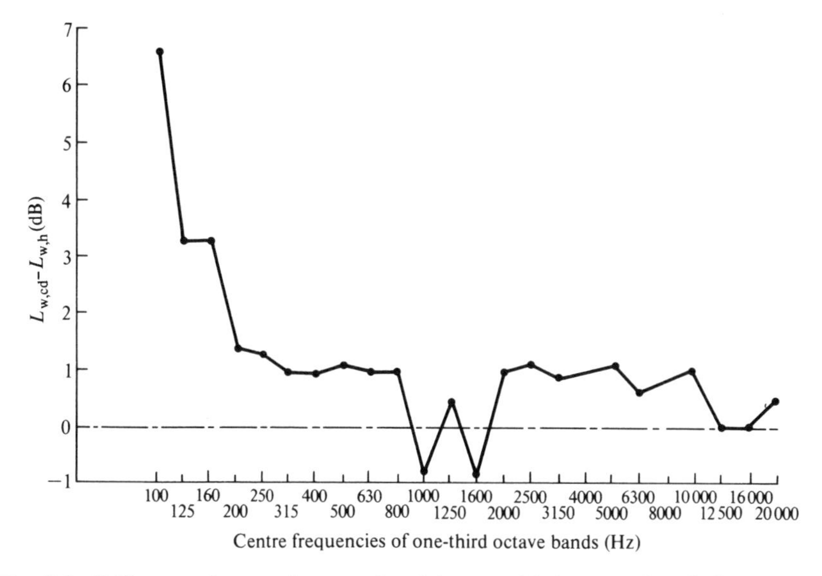

Fig. 3.9. Difference in sound-power level in one-third octave bands from measurement in semi-reverberant space $L_{w,cd}$ (nominal constant distance = 0.3 m) and from hemispherical measurement in anechoic chamber $L_{w,h}$.[3.1]

shows that for frequencies lower than 315 Hz, the differences between the sound-power level results obtained by these two measurements are not negligible.

3.3.3. *Method for reflection corrections*

A method, which requires neither a reference sound source nor the 'close-field' measurement, has been suggested[3.1] for reflection corrections in a semi-reverberant space. Measurements in one-third octave bands and A-weightings were made at 24 points on the surfaces of each of five concentric hemispheres enclosing a machine. In each case, the level of mean-square sound pressure over the surface was calculated after corrections had been made for background noise on an energy basis. The levels of the mean-square sound pressure in various frequency bands of interest were plotted against the radii of the hemispheres, for which a logarithmic scale was used. A straight line having a slope of 6 dB for each doubling of distance was then drawn tangentially to the curve. The difference between the curve and the 6-dB line at any distance is the error at that distance due to room reflections. In this way the effect of the room reflections at any distance within the range of radii tested may be determined. The error at any distance due to room reflections should be subtracted from the level of mean-square sound pressure measured at

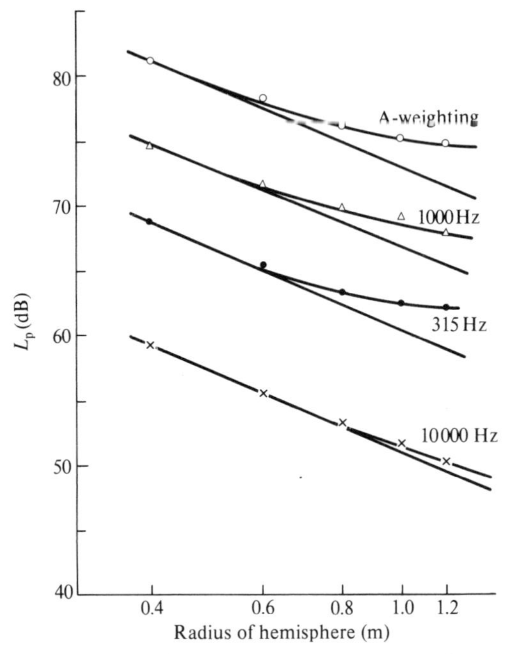

Fig. 3.10. Variation with radius of mean-square sound-pressure level over hemisphere, one-third octave bands, and A-weighting.[3.1]

that distance. Examples of the variation of the level of mean-square sound pressure in one-third octave bands with radius of the hemisphere for an electric machine in an ordinary laboratory are given in Fig. 3.10. Figure 3.11 shows the differences between sound-power levels of the machine obtained from hemispherical measurement in a semi-reverberant space without corrections and from free-field spherical measurement, the latter being regarded as the actual power levels. It is seen that the differences are much more than 1 dB except for frequencies higher than 6300 Hz. However, the differences after corrections for a reflecting floor as described in Section 4.3 and corrections for room reflections as described above are approximately within the ±1 dB range.

Although hemispheres of five different radii were used in the measurements for Fig. 3.10, it would appear that data of sufficient accuracy for most practical purposes may be obtained using only three or four distances. The method of correction is especially useful if the levels of sound power radiated by many different machines of the same size have to be measured, since no expensive calibration facility is necessary and the procedures and calculations are simple. The method should be applicable to other shapes of measuring surface if the level of the mean-square

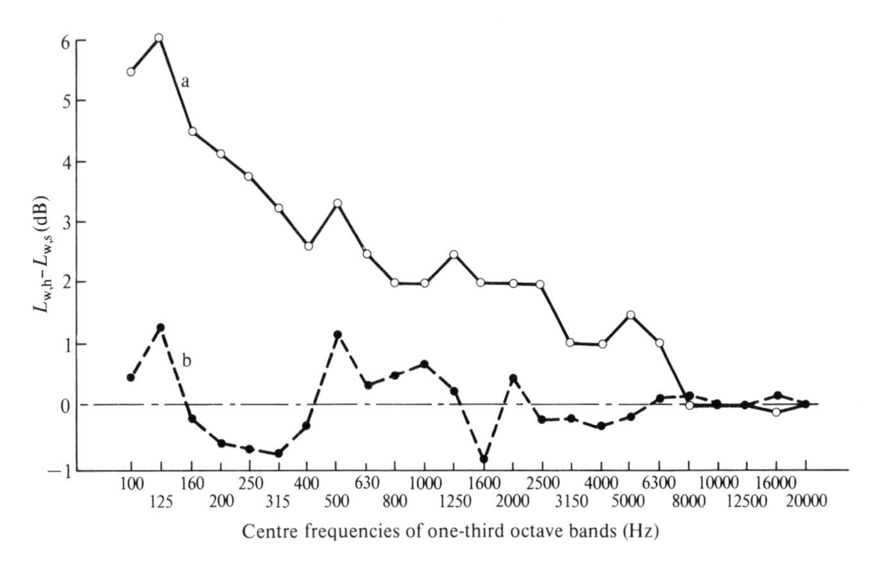

Fig. 3.11. Difference in sound-power level in one-third octave bands from hemispherical measurement in semi-reverberant space $L_{w,h}$ and from spherical measurement in anechoic chamber $L_{w,s}$.[3.1] (a) Without correction for reflections; (b) with correction for reflections.

sound pressure is plotted against the logarithms of the areas of similar measuring surfaces and a line of slope equal to 3 dB for each doubling of area is drawn tangentially to the curve.

This method has the same advantage as the reference-sound-source substitution method in that it is not necessary to express the room characteristic parameter explicitly in terms of the room constant or reverberation time.

Example 3.3

The variation with radius of the mean sound-pressure level on concentric hemispherical surfaces enclosing a machine in a semi-reverberant room for the 1000-Hz one-third octave band is shown by the 1000-Hz curve in Fig. 3.10. The sound-pressure levels on the 1.2-m radius hemispherical surface enclosing the machine, using a 12-point array, are 60, 62, 64, 65, 66, 66, 68, 68, 69, 71, 72, and 72 dB. Estimate the sound-power level of the machine, using the correction method discussed in Section 3.3.3. The effects of ground reflection can be neglected.

Solution. From eqn (1.10), the mean sound-pressure level is

$$\bar{L}_p = 10\log_{10}\left(\frac{1}{12}\sum_{i=1}^{12} 10^{0.1L_{p,i}}\right)$$
$$= 10\log_{10}\left(\tfrac{1}{12}(8.096\times 10^7)\right) = 68.3\text{ dB}.$$

From eqn (3.4), neglecting the effect of ground reflections, the sound-power level including the effect of other room reflections is

$$L_W = \bar{L}_p + 10 \log_{10}(2\pi r^2)$$
$$= 68.3 + 10 \log_{10}[2\pi(1.2)^2] = 80.9 \text{ dB}.$$

From Fig. 3.10, the contribution due to room reflections is approximately 2.3 dB at a radius of 1.2 m. The machine sound-power level in the 1000-Hz one-third octave band is therefore

$$L_W = 80.9 - 2.3 = 78.6 \text{ dB}.$$

3.3.4. *Room constant method*

If the room constant (see Appendix 1) of a semi-reverberant space is known and sound-pressure level measurements are made on a hemispherical surface enclosing the machine under test, which is placed on a hard reflecting floor, the sound-power level of the machine can be determined as

$$L_W \approx \bar{L}_p - 10 \log_{10}\left(\frac{1}{2\pi r^2} + \frac{4}{R}\right) \tag{3.9}$$

where L_W = the sound-power level in a frequency band, $\bar{L}_p$ = the mean sound-pressure level in the frequency band (determined by eqn. (1.10)), r = the radius of the hemispherical surface (m), and R = the room constant for the frequency band (m^2).

If sound-pressure level measurements are made on a spherical surface of radius r enclosing the machine under test, the machine sound-power level is given by

$$L_W \approx \bar{L}_p - 10 \log_{10}\left(\frac{1}{4\pi r^2} + \frac{4}{R}\right). \tag{3.10}$$

Equation (3.10) can be derived as follows. The sound field consists of two parts: (1) the direct field; (2) the reverberant field. In the reverberant field all waves have undergone at least one reflection from a boundary surface. The sound intensity for the direct field, assuming the directivity index to be zero and based on eqn (1.3), is given by

$$I_d = \frac{p_d^2}{\rho c} = \frac{W}{4\pi r^2} \tag{3.11}$$

where W is the sound power of the source and r the radius of the spherical measuring surface. The sound intensity for the reverberant field can be expressed as[3.15]

$$I_r = \frac{p_r^2}{\rho c} = \frac{4W}{R}. \tag{3.12}$$

The total intensity is the sum of the direct field intensity and the reverberant field intensity. Thus

$$I_t = \frac{p_t^2}{\rho c} = W\left(\frac{1}{4\pi r^2} + \frac{4}{R}\right). \qquad (3.13)$$

In practice, the total intensity varies from point to point on a spherical surface. However, the average sound intensity resulting from a number of measuring points on the surface is equal to the sound intensity at a point where the directivity index is zero; thus

$$(I_t)_{av} = \frac{(p_t)^2_{av}}{\rho c} = W\left(\frac{1}{4\pi r^2} + \frac{4}{R}\right). \qquad (3.14)$$

Based on the definition of sound-power level and sound-pressure level, eqn (3.14) becomes eqn (3.10). Equation (3.9) can be derived similarly using a hemispherical surface of $2\pi r^2$ and is useful in estimating the sound-pressure level due to a machine at a given distance from the machine (see Ex. 3.4).

Example 3.4

The room constant of a semi-reverberant space for the 1000-Hz octave band is 64 m^2. The mean sound-pressure level in the band obtained from a 1.2-m radius hemispherical surface enclosing a machine in the space is 82.0 dB.

(1) Calculate the sound-power level of the machine in the 1000-Hz octave band.

(2) Estimate the sound-pressure level in the 1000-Hz band due to the machine at a distance of 5 m from the centre of the hemispherical surface, neglecting directivity. The centre of the machine base is at the centre of the hemispherical surfacc.

Solution. From eqn (3.9), the sound-power level of the machine in the 1000-Hz octave band is

$$\begin{aligned} L_W &= \bar{L}_p - 10\log_{10}\left(\frac{1}{2\pi r^2} + \frac{4}{R}\right) \\ &= 82.0 - 10\log_{10}\left(\frac{1}{2\pi(1.2)^2} + \frac{4}{64}\right) \\ &= 82.0 + 7.6 = 89.6\ \text{dB} \quad (\text{re } 1\times 10^{-12}\ \text{W}). \end{aligned}$$

Neglecting directivity, the sound-pressure level at a distance of 5 m from the centre of the hemispherical surface is the same as the mean sound-pressure level on the 5-m radius hemispherical surface. Thus, from eqn (3.9) the sound-pressure level required is given by

$$\begin{aligned}\bar{L}_p &= L_W + 10\log_{10}\left(\frac{1}{2\pi r^2}+\frac{4}{R}\right)\\ &= 89.6 + 10\log_{10}\left(\frac{1}{2\pi(5)^2}+\frac{4}{64}\right)\\ &= 89.6 - 11.6 = 78\text{ dB} \quad (\text{re } 2\times10^{-5}\text{ N m}^{-2}).\end{aligned}$$

3.3.5. *Oil Companies Materials Association methods*[3.17]

In petroleum and chemical plants, the use of special anechoic or reverberant chambers for measuring machinery noise is not always practicable because equipment and machines are bulky. Furthermore many machines have to be tested indoors *in situ*. The Oil Companies Materials Association has specified the following methods for measuring machinery noise.

3.3.5.1. *Small-source method.* This *small-source method* is to be used for noise sources whose linear dimensions are small in relation to the radius of the measuring surface. The measuring surface is a hemisphere terminating on the ground. The radius of the hemisphere is at least twice the typical source dimension D_0, defined as

$$D_0 = [(l_1/2)^2 + (l_2/2)^2 + l_3^2]^{1/2} \tag{3.15}$$

where l_1, l_2, and l_3 are the length, width, and height of the reference surface, respectively. The radius shall be such that the hemisphere is at least 1 m from the reference surface. The microphone positions are situated on the measuring surface (see Fig. 3.12). In order to make a correction for the test environment, a two-surface method suggested by Diehl[3.19] should be used. Thus, a second set of measurements should be made on a second hemispherical measuring surface of area approximately four times that of the first measuring surface. The estimate of the sound-power level based on the larger surface is

$$L'_{Wa} = \bar{L}'_p + 10\log_{10}S' \tag{3.16}$$

and that based on the smaller surface is

$$L_{Wa} = \bar{L}_p + 10\log_{10}S \tag{3.17}$$

where $\bar{L}'_p$ and $\bar{L}_p$ are the mean sound-pressure levels and S' and S are the surface areas for the larger and smaller surfaces, respectively. If L'_{Wa} exceeds L_{Wa} by more than 2 dB, an environmental correction E_2 should be determined from Fig. 3.13 and shall be applied to L_{Wa}.

Therefore, the sound-power level of the source is given by

$$L_W = \bar{L}_p + 10\log_{10}S - E_2. \tag{3.18}$$

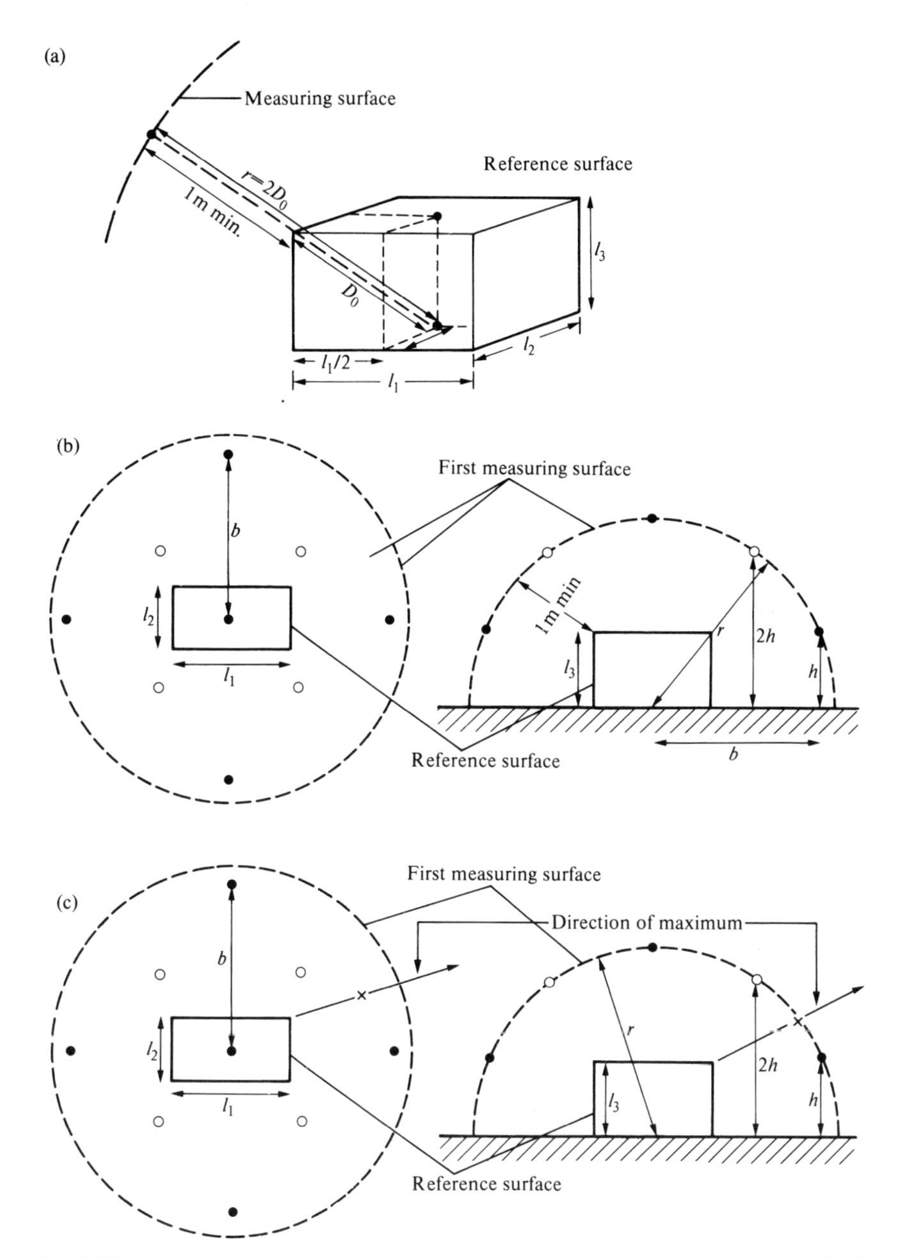

Fig. 3.12. Measuring surface and measuring points for small-source method.[3.17] (a) Determination of radius r of measuring surface. Minimum radius of measuring surface $r = 2D_0 = 2\sqrt{[(l_1/2)^2 + (l_2/2)^2 + l_3^2]}$; (b) measuring points. (●) Measuring points with overhead point. Preferred positions: $h = 0.4\,r$, $b = 0.9\,r$; (○) Additional measuring points to replace overhead point; (c) small-source method: measuring points for source with directivity. (●) Basic measuring points. Preferred positions: $h = 0.4\,r$, $b = 0.9r$. (○) Additional measuring points. (×) Point to measure maximum.

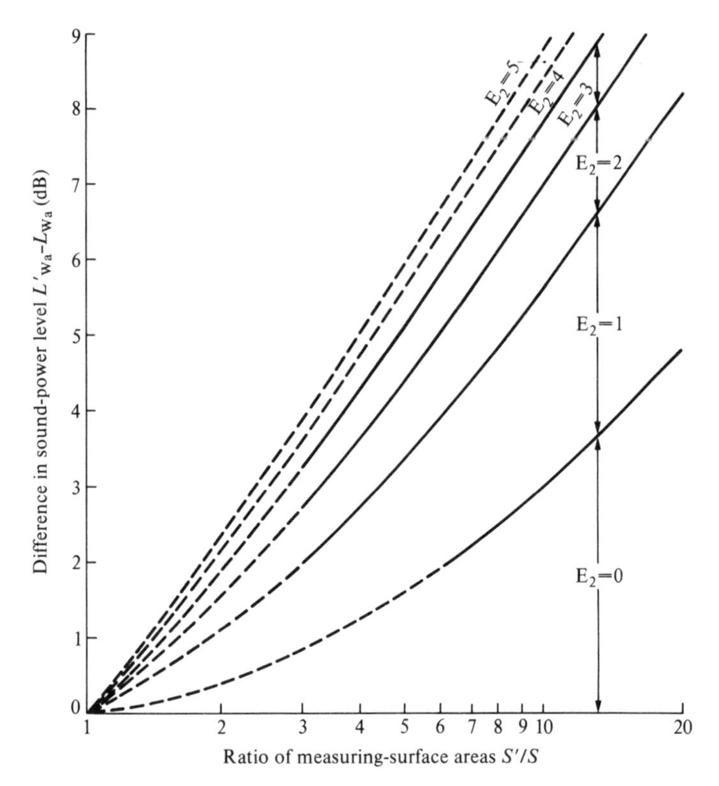

Fig. 3.13. Correction for semi-reverberant test environment (two-surface method).[3.17]

If the difference between L'_{Wa} and L_{Wa} is 2 dB or less, the sound-power level based on the smaller surface shall be taken as the sound power level.

3.3.5.2. *Large-source method.* This method shall be used for sources whose dimensions are large relative to the measuring distance, but only when circumstances do not allow the use of the small-source method. The shape of the measuring surface should preferably be a conformal surface which envelopes the reference surface box (see Fig. 3.14). The preferred and minimum measuring distance is 1 m from the reference surface. The sound-power level is calculated from

$$L_W = (\bar{L}_p - E_1) + 10\log_{10}S - E_2 \tag{3.19}$$

where E_1 is the near-field correction and E_2 is the environment correction. The near-field correction E_1 depends on the angle subtended at the microphone by the source surface. For convenience, it can be determined

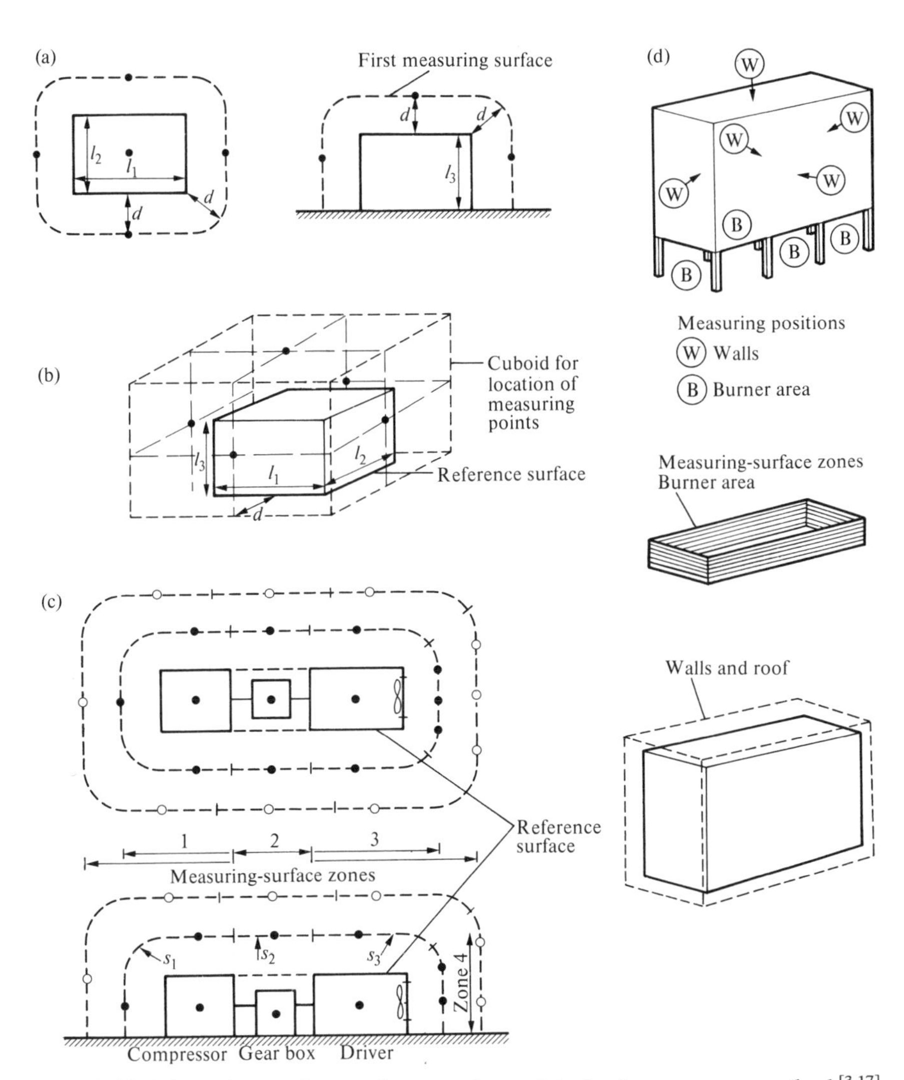

Fig. 3.14. Measuring surface and measuring points for large-source method.[3.17] For source with evenly distributed radiation: (a) conformal surface; (b) measuring points; (c) source requiring four measuring-surface zones, typified by compressor set with gearbox (with acoustic lagging on piping). (●) Measuring points—inner surfaces. (○) Measuring points—outer surfaces. Note: the measuring-surface zones on the outer surface may not be similar to those on the inner surface. For source with unequal radiating surfaces: (d) measuring positions for a floor-fired furnace.

from the value of Q, which is defined as

$$Q = \frac{\text{area of the reference surface}}{\text{area of the measuring surface}}.$$

The value of the near-field correction E_1 can be obtained from the following table.

Q	E_1 (dB)
$0.9 < Q < 1$	3
$0.7 < Q < 0.9$	2
$0.4 < Q < 0.7$	1
$0.0 < Q < 0.4$	0

The environmental correction E_2 can be determined by the two-surface method described in the small-source method. However, in the determination of the estimate of the sound-power level, the near-field correction should be included. Thus, based on the smaller surface, the estimate of the sound-power level is

$$L_{\mathrm{Wa}} = \bar{L}_{\mathrm{p}} - E_1 + 10 \log_{10} S. \tag{3.20}$$

Based on the larger surface, the estimate of the sound-power level is

$$L'_{\mathrm{Wa}} = \bar{L}'_{\mathrm{p}} - E'_1 + 10 \log_{10} S'. \tag{3.21}$$

Alternatively, if the room constant R (see Appendix 1) is known, the environmental correction factor E_2 can be determined by

$$E_2 = 10 \log_{10}(1 + 4S/R). \tag{3.22}$$

The above equation is plotted in Fig. 3.15. For values of E_2 larger than 3 dB, the measurement surface is in the reverberant field and the present correction procedure is not valid. If E_2 is larger than 3 dB, the measurement surface should be chosen closer to the machine or sound absorbing material should be added temporarily in the test space to increase the ratio of the room constant to the surface area of the measuring surface R/S.

3.3.5.3. *Composite and linear sources.* When a source consists of several well-defined components, it may be necessary to obtain separate measurements of the components. Some equipment such as pumps and compressors has associated pipework and the noise emitted by the pipework may be greater than that emitted by the equipment. Pumps and compressors should therefore be treated as three separate sources comprising the

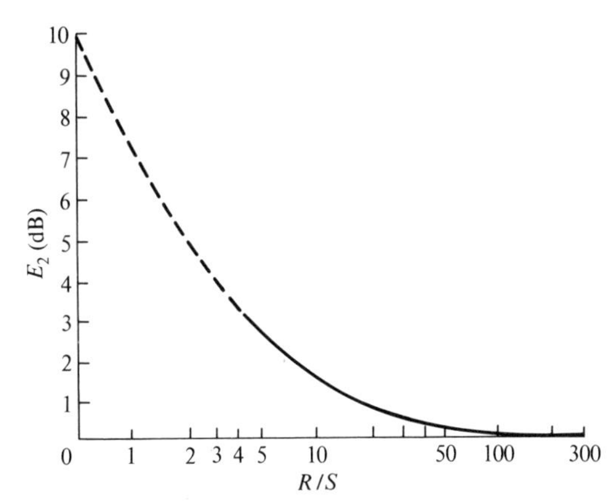

Fig. 3.15. Correction for semi-reverberant test environment (room constant method).[3.17]

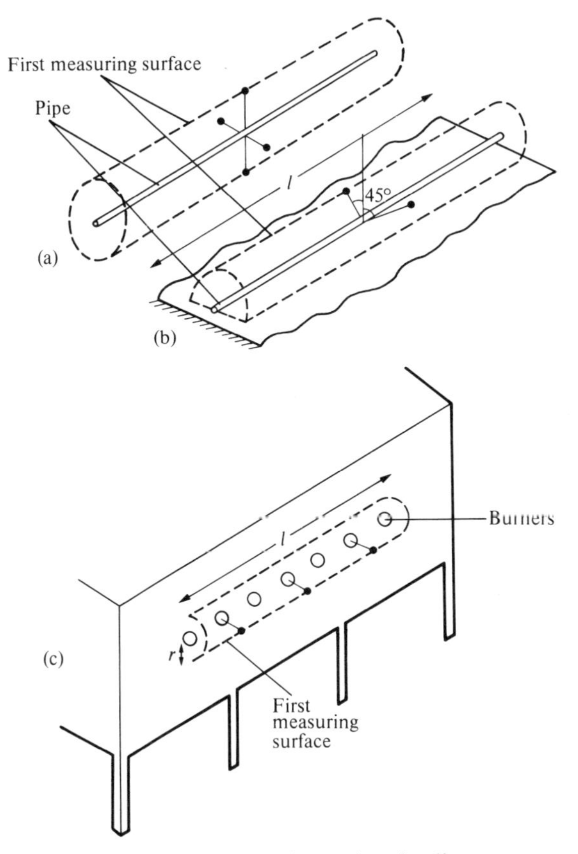

Fig. 3.16. Measuring surface and measuring points for linear sources.[3.17] Typical measuring positions: (a) Overhead pipe (no reflecting surface), $S = 2\pi r$; (b) pipe near reflecting surface, $S = \pi r$; (c) row of burners in wall-fired furnace (wall acting as reflecting surface). (●) Measuring positions.

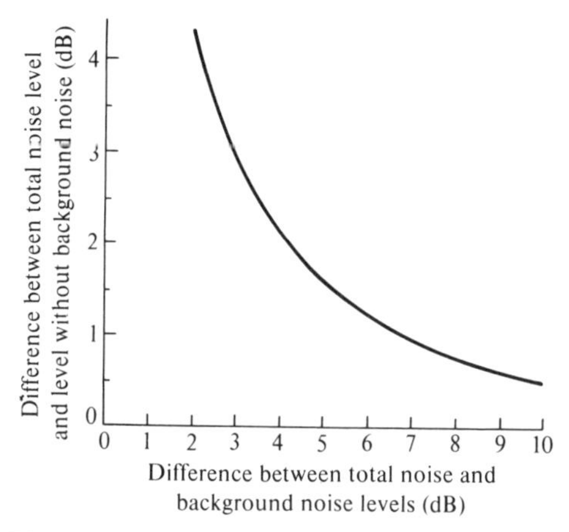

Fig. 3.17. Correction for background noise.

suction piping, the machine and the discharge piping. The sound-power level of the piping is that of the complete run from the equipment to the next plant item. Typical measuring positions for a pipe are shown in Fig. 3.16.

3.3.6. *Corrections for background noise*

A common difficulty associated with measurements under semi-reverberant conditions is the considerable background noise which is often present. If the sound-pressure level measurement in any frequency band exceeds the background sound-pressure level in the band by more than 10 dB, no correction is normally necessary. If the difference is less than 3 dB, correction for background noise is very uncertain and accurate measurement results cannot be obtained. Between these two values, a correction using Fig. 3.17 should be made.

Example 3.5

In a semi-reverberant space the sound-pressure level measured at a given point around a machine in the 500-Hz octave band is 62.0 dB. When the machine is switched off, the background-noise sound-pressure level at the same point in the 500-Hz band is 57.0 dB. Determine the sound-pressure level due to the machine alone at the given point without the background noise. It can be assumed that the other noise sources emit steady noise and their operating conditions remain unchanged.

Solution. The difference between the total noise sound-pressure level and the background sound-pressure level in the 500-Hz octave band is

$$L_{\mathrm{p,total}} - L_{\mathrm{p,background}} = 62 - 57 = 5\ \mathrm{dB}.$$

From Fig. 3.17, the difference between the total noise sound-pressure level and the sound-pressure level without background noise corresponding to a 5-dB difference between the total noise and background noise levels is found to be 1.6 dB. Thus, the sound-pressure level at the given point due to the machine alone is

$$L_{p,machine} = L_{p,total} - 1.6$$
$$= 62.0 - 1.6 = 60.4 \text{ dB} \approx 60 \text{ dB}.$$

3.4. Measurement in reverberant rooms

Machinery noise measurements can be made in a reverberant room, which is a room having highly sound-reflecting, non-parallel walls, ceiling, and floor. Ideally, as the sound from a source is reflected back and forth repeatedly between the highly reflecting surfaces of the room, these multiple reflections would create a completely diffuse sound field, i.e. a sound field having equal energy density at all points. However, a completely diffuse sound field does not exist in reality.

The main advantages of using a reverberant room are

(1) It is considerably less expensive to build a reverberant room than an anechoic room;

(2) A reverberant room is appropriate for the reference sound-source substitution method which determines the approximate sound-power level of a noise source by comparing the sound-power level of the source in the room with that of a reference sound source of known sound-power level in the same room.

The main disadvantages of reverberant room measurements are that the method cannot measure the directivity of a noise source and the method gives less accurate sound-power results, especially for noise sources having predominant pure-tone components. The reasons for the inaccuracy will be discussed below.

Owing to reflections, interference patterns near each boundary wall are set up in a reverberant room. Assuming sound-pressure waves of wavelength λ and unit amplitude strike a wall from all directions, it can be shown[3.7] that the mean-square sound pressure near the wall is given by

$$\overline{p_x^2} = 1 + \sin\left(\frac{4\pi x}{\lambda}\right) \Big/ \left(\frac{4\pi x}{\lambda}\right) \tag{3.23}$$

where x = distance from the wall (m) and λ = wavelength of the sound wave (m).

Figure 3.18 gives the variations of $\overline{p_x^2}$ with $4\pi x/\lambda$, based on eqn

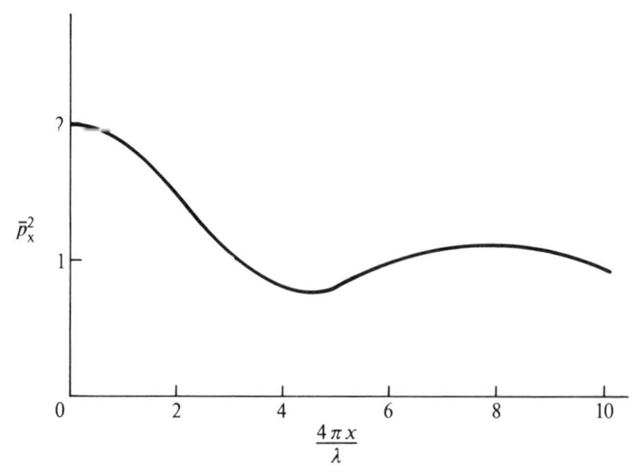

Fig. 3.18. Variation of mean-square sound pressure near a reflecting wall. x = distance from wall; λ = sound wavelength.

(3.23). It is seen that the mean-square sound pressure doubles at the wall surface and fluctuates with decreasing amplitude. There are more pronounced fluctuations in sound pressure in the vicinity of a two-wall edge and a three-wall corner.[3.7]

Since many machines radiate one or more predominant pure-tone noise components which would produce complicated interference patterns in a reverberant room, the sound-power output of these machines in a reverberant room is not the same as that in a free field. In order to minimize the effect of interference on sound pressure and sound power, ISO 3741[3.8] recommends that all parts of the machine under test should be at least 1.5 m from any wall of the room. All microphone positions or any microphone traverse shall not lie in any plane within 10° of a room surface. Neither any microphones nor any point on the traverse shall be closer than $\lambda/2$ to any room surface, where λ is the wavelength of the lowest frequency of interest. The volume of the noise source is preferably less than 1 per cent of the volume of the reverberation room.

To determine the approximate sound power emitted by a machine in a reverberant room, it is assumed that the sound energy density E which is defined as the average of the acoustic energy per unit volume of the air particles, is uniform throughout the room. Since the speed of sound energy propagation is c (m s^{-1}), the sound intensity is given by

$$I = Ec \quad \mathrm{W\,m^{-2}}. \tag{3.24}$$

Taking $I = p^2/\rho c$ as in eqn (1.3), we have

$$E = \frac{p^2}{\rho c^2} \quad \mathrm{W\,s\,m^{-3}}. \tag{3.25}$$

Equation (3.25) can also be derived directly from the potential energy and kinetic energy of the air particles.[3.9]

The total sound power impinging on a unit area of wall in a reverberant room with a uniform sound field can be shown to be[3.10]

$$I_{\text{imp}} = \frac{cE}{4} \quad \text{W m}^{-2}. \tag{3.26}$$

If the average absorption coefficient of the room is $\bar{\alpha}$, the sound power absorbed by the room is

$$P_{\text{abs}} = \frac{cE}{4} S\bar{\alpha} \quad \text{W} \tag{3.27}$$

where S is the total surface area of the room in m^2.

If a sound source emits steady noise and the room has reached its steady-state condition with a steady-state energy density $E_{\text{s,s}}$, then the sound power from the source must be equal to the sound power absorbed by the room. Thus, the sound power of the source is

$$W = \frac{cE_{\text{s,s}}}{4} S\bar{\alpha}. \tag{3.28}$$

The average absorption coefficient $\bar{\alpha}$ is related to the reverberation time (see Appendix 1) as[3.14]

$$T = 0.161 \frac{V}{S\bar{\alpha}} \tag{3.29}$$

where V is the total volume of the room in m^3 and T is the reverberation time of the room in seconds (see Appendix 1 for its determination). Combining eqns (3.25), (3.28), and (3.29), we have

$$W = \frac{p_{\text{s,s}}^2}{4\rho c}\left(0.161 \frac{V}{T}\right).$$

The sound-power level of the source is thus equal to

$$\begin{aligned} L_{\text{W}} = 10 \log \frac{W}{W_{\text{ref}}} &= 10 \log_{10} \frac{p_{\text{s,s}}^2}{p_{\text{ref}}^2} + 10 \log_{10}\left(\frac{V}{T}\right) + 10 \log_{10}\left(\frac{0.161 p_{\text{ref}}^2}{4\rho c W_{\text{ref}}}\right) \\ &= L_{\text{p,s,s}} + 10 \log_{10} V - 10 \log_{10} T - 14 \text{ dB}. \end{aligned} \tag{3.30}$$

Since the sound-energy density and the sound-pressure level in an actual reverberant room vary from point to point, it is necessary to carry out a considerable number of sound-pressure level measurements in order to obtain the estimate of $L_{\text{p,s,s}}$. Baade[3.11] mentioned that nearly 50 measurements at locations which are at least one-half wavelength apart are required to achieve a 90 per cent confidence level that the error in $\bar{L}_{p,s,s}$ does not exceed 1 dB. In practice, a travelling microphone system[3.12] can be used to obtain an approximate $\bar{L}_{p,s,s}$ value. For broad-band sources, ISO 3741[3.8] specfies that a microphone should traverse at constant speed a path of at least 3 m in length while the microphone signal is being averaged on a mean-square basis. Alternatively, at leat three fixed microphones or microphone positions should be used, placed at a distance of $\frac{1}{2}\lambda$ from each other, where λ is the wavelength of the lowest frequency of interest. For discrete frequency and/or narrow-band noise sources, six or more microphone positions should be used and tests may be required for a number of source locations. Readers can refer to ISO 3742[3.20] for the required number of microphone positions and source locations. Using the measured mean sound-pressure level $\bar{L}_{p,s,s}$ value instead of the $L_{p,s,s}$ value, eqn (3.30) becomes

$$L_W = \bar{L}_{p,s,s} + 10\log_{10}V - 10\log_{10}T - 14\text{ dB}. \tag{3.31}$$

3.5. Averaging sound-pressure levels

The use of eqn (1.10) to find the mean sound-pressure level involves taking antilogarithms of the sound-pressure levels, averaging, and then taking logarithms again and is rather time-consuming. However, if the sound-pressure levels of a set of measurements have a spread (the maximum variation of L_p values) of less than 5 dB, it is common practice to assume that, to a sufficient degree of accuracy, the mean sound-pressure level is equal to the arithmetic average of the sound-pressure levels $L_{p,av}$. Thus, the mean sound-pressure level is

$$\bar{L}_p \approx L_{p,av} = \frac{1}{n}\sum_{i=1}^{n} L_{p,i} \quad \text{for a spread} \leqslant 5\text{ dB}. \tag{3.32}$$

Cox[3.13] showed that the maximum error between $\bar{L}_p$ and $L_{p,av}$, assuming that a proportion η of the sound-pressure levels is at the upper limit of the spread, and a proportion $(1-\eta)$ is at the lower limit, is given by

$$(\bar{L}_p - L_{p,av})_{max} = 10\log_{10}\left(\frac{\ln K}{K-1}\,K^{[(1/\ln K)-1/(K-1)]}\right) \tag{3.33}$$

where $K = 10^{S/10}$ and S is the spread.

Based on eqn (3.33) the maximum error in the arithmetic average $L_{p,av}$ is approximately 0.7 dB for a spread of 5 dB. From over 250 sets of noise data measured from electric machines, it has been found[3.1] that the distribution of sound-pressure levels around an electric machine on

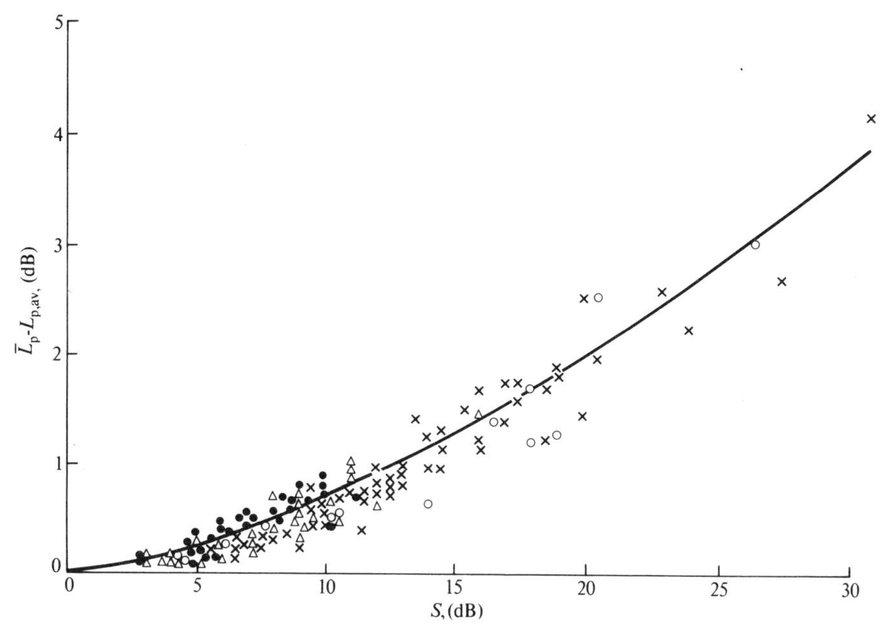

Fig. 3.19. Correction for arithmetic-average sound-pressure level.[3.1] $\bar{L}_p$ = mean-square sound-pressure level; $L_{p,av}$ = arithmetic-average sound-pressure level; S = spread in dB. —— $\bar{L}_p - L_{p,av} = 0.0184S^{1.56}$. (×) motor 1; (○) motor 2; (△) motor 3; (●) motor 4.

various spherical and hemispherical measuring surfaces in a semi-reverberant space and in an anechoic chamber is approximately of normal (Gaussian) distribution and that the difference between the mean sound-pressure level $\bar{L}_p$ and the linear arithmetic average of sound-pressure levels $L_{p,av}$ can be approximately expressed as a function of the spread S as

$$\bar{L}_p - L_{p,av} = 0.0184S^{1.56}. \tag{3.34}$$

Equation (3.34) is plotted in Fig. 3.19 for spreads up to 30 dB and can be used to determine the approximate mean sound-pressure level $\bar{L}_p$ value very quickly from the arithmetic average $L_{p,av}$ and the spread S value. Though eqn (3.34) and Fig. 3.19 were derived from noise measurements on electric machines, they can be used to estimate the approximate $\bar{L}_p$ value for sound-pressure level data obtained from other machines which have a sound-pressure level distribution of approximately Gaussian type.

Example 3.6

Determine the approximate mean sound-pressure level for the sound-pressure level data given in Example 3.1 using the arithmetic average of the sound-pressure levels and the approximate correction shown in Fig. 3.19.

Solution. The arithmetic average of sound-pressure levels for the data in Example 3.1 is

$$L_{p,av} = \frac{1}{12} \sum_{i=1}^{12} L_i = \frac{1}{12}(76+78+79+82+83+86+88+88$$
$$+89+89+90+92)$$
$$= 85.0 \text{ dB}.$$

The spread for the sound-pressure levels is

$$S = 92-76 = 16 \text{ dB}.$$

From Fig. 3.19 for a spread of 16 dB, we have

$$\bar{L}_p - L_{p,av} = 1.4 \text{ dB}.$$

Thus the approximate mean sound-pressure level is

$$\bar{L}_p = L_{p,av} + 1.4 = 85.0 + 1.4 = 86.4 \text{ dB} \approx 86 \text{ dB}.$$

References

[3.1] Ellison, A. J., Moore, C. J., and Yang, S. J. (1969). Methods of measurement of acoustic noise radiated by an electric machine. *Proc. IEE* **116,** 1419–31.

[3.2] Yang, S. J. (1975). Acoustic noise from small 2-pole single-phase electric machines. *Proc. IEE* **122,** 1391–6.

[3.3] British Standards Institution (1981). *Engineering methods for determination of sound power levels for sources in free-field conditions over a reflective plane,* BS 4196, Part 4. British Standards Institution, London.

[3.4] ISO (1970). *Test code for the measurement of airborne noise emitted by rotating electrical machines,* ISO/R 1680. ISO, Geneva.

[3.5] BEAMA (1967). *BEAMA recommendations for the measurement and classification of acoustic noise from rotating electrical machines,* BEAMA Publication 225. BEAMA, London.

[3.6] King, A. J. (1964). Setting standards for machine noise. *Engineering* **198,** 93–5.

[3.7] Waterhouse, R. V. (1955). Interference patterns in reverberant sound fields. *J. acoust. Soc. Am.* **27,** 247–58.

[3.8] ISO (1975). *Determination of sound power levels of noise sources—precision methods for broad-band sources in reverberation rooms,* ISO 3741. ISO, Geneva.

[3.9] Rschervkin, S. N. (1963). *The theory of sound.* Pergamon, Oxford.

[3.10] Kinsler, L. E. and Frey, A. R. (1962). *Fundamentals of acoustics,* 2nd edn. John Wiley, New York.

[3.11] Baade, P. K. (1968). Sound power measurement in reverberant rooms.

Symposium of 76th meeting of Acoust. Soc. Amer., Acoust. Soc. Amer., Cleveland.

[3.12] Ploner, B. (1967). Determining the sound power of rotating electrical machines in a reverberation room. *Brown Boveri Rev.* **54,** 648–57.

[3.13] Cox, H. (1966). Linear versus logarithmic averaging. *J. acoust. Soc. Am.* **39,** 688–90.

[3.14] Beranek, L. L. (1971). *Noise and vibration control.* McGraw-Hill, New York.

[3.15] Fricke, F. R. and Tree, D. R. (1975). Room acoustics. In *Reduction of machine noise* (ed. M. J. Crocker) pp. 12–26. Purdue University Press, Purdue, Indiana.

[3.16] ISO (1977). *Determination of sound power levels of noise sources—precision methods for anechoic and semi-anechoic rooms,* ISO 3745. ISO Geneva.

[3.17] OCMA (1980). *Noise procedure specification,* OCMA Specification No. NWG1 (Rev. 2). Institute of Petroleum, London.

[3.18]. ISO (1981). *Engineering methods for free field conditions over a reflecting plane,* ISO 3744. ISO, Geneva.

[3.19] Diehl, G. M. (1977). Sound power measurements on large machinery installed indoors; two-surface method. *J. acoust. Soc. Am.* **61**(2), 449–55.

[3.20] ISO (1975). *Precision methods for discrete-frequency and narrow-band sources in reverberation rooms,* ISO 3742. ISO, Geneva.

Further reading for Chapter 3

[1] ISO (1979). *Determination of sound power levels of noise sources—survey method,* ISO 3746. ISO, Geneva.

[2] ISO (1982). *Survey method using a reference sound source,* ISO/DP 3747. ISO, Geneva.

[3] ISO (1982). *Engineering method for small, omnidirectional sources under free field conditions over a reflecting plane,* ISO/DP 3748. ISO, Geneva.

[4] Shew, A. T. (1980). Improvement of sound power level nearfield measurement on electrical machines in situ. *Proceedings of Inter-Noise 80,* pp. 1071–8. Noise Control Foundation, New York.

[5] Yousri Gerges, S. N. and De Araujo, M. A. N. (1980). Acoustic power contribution due to source near the reverberation chamber boundaries. *Proceedings of Inter-Noise 80,* pp. 1079–82. Noise Control Foundation, New York.

[6] Woehrle, K. K. (1980). Impulsive noise: determination of sound power using reverberation room methods. *Proceedings of Inter-Noise 80,* pp. 1087–92. Noise Control Foundation, New York.

[7] Maling, G. C. and Clark, M. G. (1981). Sound power measurements in reverberation rooms. *Proceedings of Inter-Noise 81,* pp. 921–4. Nederlands Akoestisch Genootschap, Delft.

[8] Tichy, J. (1981). Precision of the sound power determination in a reverberant room with rotating diffuser. *Proceedings of Inter-Noise 81,* pp. 925–8. Nederlands Akoestisch Genootschap, Delft.

[9] Brüel, P. V. (1981). Sound power measurements of household appliances. *Proceedings of Inter-Noise 81,* pp. 917–20. Nederlands Akoestisch Genootschap, Delft.

[10] OCMA (1980). *Guide to the use of noise procedure specification NWG1,* OCMA Publication No. NWG3 (Rev. 2). Institute of Petroleum, London.

[11] OCMA (1980). *General specification for silencers and acoustic enclosures*, OCMA Specification No. NWG4. Institute of Petroleum, London.

[12] ISO (1980). Acoustics—determination of sound power sources—guidelines for the use of basic standards and for the preparation of noise test codes, ISO 3740. ISO, Geneva.

[13] ISO (1976). *Engineering methods for special reverberation test rooms*, ISO 3743. ISO, Geneva.

[14] ISO (1975). *Measurement of reverberation time in auditoria*, ISO 3382. ISO, Geneva.

[15] Upton, R. (1977). Automated measurements of reverberation time using the digital frequency analyser type 2131. *B & K tech. Rev.* **2,** 3–18.

[16] Schultz, T. J. (1973). Outlook for in-situ measurement of noise from machines. *J. acoust. Soc. Am.* **54**(4), 982–4.

[17] Lubman, D., Waterhouse, R. V., and Chen, C. S. (1973). Effectiveness of continuous spatial averaging in a diffuse sound field. *J. acoust. Soc. Am.* **53**(2), 650–9.

[18] Ebbing, C. E. and Maling, J. G. C. (1973). Reverberation room qualification for determination of sound power of sources of discrete-frequency of sound. *J. acoust. Soc. Am.* **54**(4), 935–49.

[19] Holmex, C. I. (1977). Investigation of procedures for estimation of sound power in the free field above a reflecting place. *J. acoust. Soc. Am.* **61**(2), 465–75.

[20] Hübner, G. (1977). Qualification procedures for free-field conditions for sound power determination of sound sources and methods for the determination of the appropriate environmental correction. *J. acoust. Soc. Am.* **61**(2), 456–64.

4. Accuracy in sound-power results

Since the noise field of a machine varies from point to point and changes with its acoustic environment, there are a number of errors involved in the sound-power results obtained by the sound-pressure measurement methods discussed in Chapter 3. We shall discuss the errors due to

(1) Finite number of measuring points;
(2) The 'far-field' assumption;
(3) Ground reflection.

Other problems, e.g. electrical noise floor, wind noise, and microphone alignment, will also be discussed.

4.1. Error due to finite number of measuring points

Using a given measuring surface enclosing a machine in a free-field condition, the true mean sound-pressure level on the surface can be obtained only when a very large number of measuring points are taken on the surface since the sound-pressure level values on the surface vary from point to point. However, in practice, a limited number of measuring points are taken and it is important, therefore, to estimate the error due to the finite number of measuring points.

Assuming the distribution of sound-pressure level values over the measuring surface to be of approximately the normal (Gaussian) type, the difference between the true and measured values of $(p_i^2)_{av}/p_{ref}^2$ at a certain confidence level can be expressed (see Appendix 2) by[4.1]

$$\left|\left(\frac{(p_i^2)_{av}}{p_{ref}^2}\right)_{true} - \left(\frac{(p_i^2)_{av}}{p_{ref}^2}\right)_{measured}\right| = \frac{t\sigma_E}{\sqrt{(n-1)}} \tag{4.1}$$

where t is Student's t for $(n-1)$ degrees of freedom, σ_E is the standard deviation of the p_i^2/p_{ref}^2 values, and n is the number of measuring points. Thus the error in sound-power level, which is taken as the difference between the upper limit of the true sound-power level for a confidence level of α per cent and the sound-power level calculated from n sound-pressure level measurements, is given by

$$\begin{aligned} L_{W_{true}} - L_{W_{measured}} &= 10\log_{10}\left(\frac{[(p_i^2)_{av}/p_{ref}^2]_{true}}{[(p_i^2)_{av}/p_{ref}^2]_{measured}}\right) \\ &= 10\log_{10}\left(\frac{[(p_i^2)_{av}/p_{ref}^2]_{measured} + t\sigma_E/\sqrt{(n-1)}}{[(p_i^2)_{av}/p_{ref}^2]_{measured}}\right) \\ &= 10\log_{10}\left[1 + n\sigma_E t\Big/\left(\sqrt{(n-1)}\sum_{i=1}^{n} 10^{L_i/10}\right)\right] \text{ dB} \end{aligned} \tag{4.2}$$

where t is the critical value of Student's t for $(n-1)$ degrees of freedom which will be exceeded with the probability given in Table A.2.1 in Appendix 2.

From the analyses of noise data obtained from spherical measurements on four different electric machines in free-field and in semi-reverberant space with A-weighting and various octave bands, one-third octave bands, and 6 per cent narrow bands, it has been found[4.2] that the standard deviation of the p_i^2/p_{ref}^2 values σ_E is related to the spread S, the difference in dB between the highest and lowest sound-pressure levels, by

$$\sigma_E = 0.087 S^{0.82} \sum_{i=1}^{n} 10^{L_{p,i}/10}/n. \tag{4.3}$$

Combining eqns (4.2) and (4.3) yields the error in the power level

$$L_{W_{true}} - L_{W_{measured}} = 10 \log_{10}[1 + 0.087 t S^{0.82}/\sqrt{(n-1)}]\ \mathrm{dB}. \tag{4.4}$$

Equation (4.4) is plotted in Fig. 4.1, which gives the sound-power level error at a 97.5 per cent confidence level for spreads of up to 30 dB and measuring points up to 300.

The variation in sound-power level for an electric machine obtained with different numbers of measuring points is shown in Table 4.1, where the experimental results from spherical measurements with various numbers of measuring points up to 137 for four different radii in an anechoic

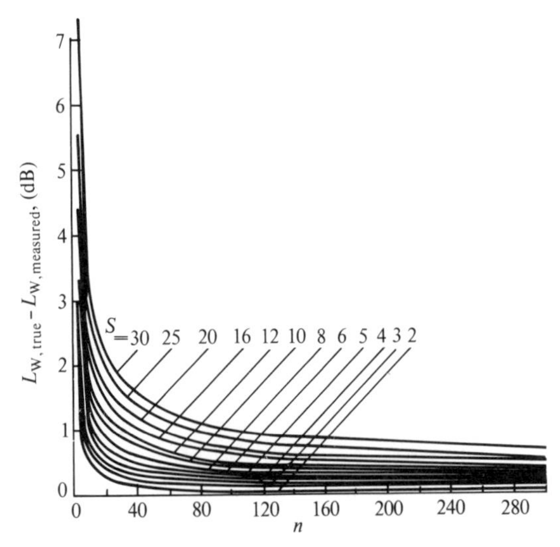

Fig. 4.1. Variation of error in sound-power level with number of measuring points and spread of sound-pressure levels (97.5 per cent confidence level).[4.2] S = spread of sound-pressure levels in one set of measurements; n = number of measuring points.

TABLE 4.1. Variation in sound power obtained with diffeent numbers of measuring points, based on data from four different machines[4.2]

Frequency (Hz) (centre of narrow-bands)	Spread of s.p.l.s (dB)	Maximum variation from value obtained using 272 points (dB). M: measured; C: calculated									
		137 pts		92 pts		36 pts		20 pts		8 pts	
		M	C*	M	C*	M	C*	M	C*	M	C*
100	31	0.1	0.3	0.1	0.5	0.8	1.4	0.5	2.4	1.3	4.6
600, 880, 2200, 2800, 3500	14	0.1	0.2	0.1	0.3	0.8	0.9	0.8	1.3	1.5	2.3
1100, 1200, 1400, 4400, 5500, 7700	20	0.1	0.2	0.2	0.4	0.4	1.1	0.8	1.9	3.3	3.8
A-weighting	7	0.1	0.1	0.2	0.1	0.4	0.4	0.3	0.6	0.5	1.1

* Calculated at 97.5 per cent confidence level.

chamber are compared with those obtained by the statistical method described here. It is seen that the variation in the sound-power level figures produced by a reduction in the number of measuring points is within the error obtained from eqn (4.4) at 97.5 per cent confidence level. Equation (4.4) can be used to find the error in sound-power level due to a finite number of measuring points for other shapes of measuring surface. This follows since the experimental relationship shown in eqn (4.3) was found to be valid not only for spherical measurements in free field but also for a semi-reverberant space and for other shapes of measuring surface, the results obtained from four electric machines under these conditions having been analysed. Equations (4.2) or (4.4) enable one to estimate whether the number of points employed in a noise measurement is satisfactory for a given accuracy.

The International Standard ISO 3745[4.3] suggests that, for discrete microphone arrays, it is desirable to start with 10 microphone positions on a hemispherical measuring surface or 20 positions on a spherical surface and that the number of measurement points is regarded as sufficient for what is termed a *Precision method* if the difference in dB between the highest and lowest sound-pressure levels, i.e. the spread in sound-pressure levels, measured on the measuring surface in any frequency band of interest, is numerically less than half the number of measurement points. If this requirement is not met, measurements shall be made at additional measurement positions. This recommendation provides a simple but rough guide on the choice of measurement points. Hübner *et al.*[4.4] investigated this recommendation and concluded that

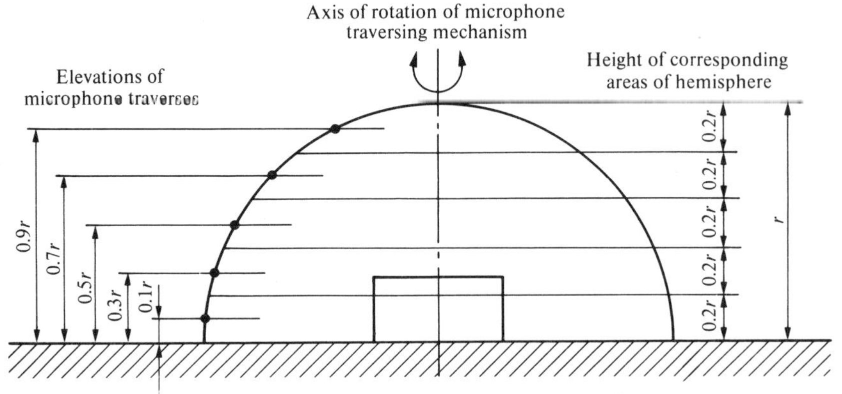

Fig. 4.2. Parallel circular paths for microphone traverses in a free field over a reflecting plane (Reproduced by permission of the British Standards Institution, London).[4.3]

the number of measuring points should be greater than the ISO 3745 recommended values, if accurate sound-power level is required. Alternatively, the use of sound measuring equipment capable of integrating and averaging the square of the sound-pressure output of a slow-moving microphone which travelled along a given measuring path, would increase the effective number of measuring points to a very large value and hence improve the accuracy in sound-power results.[4.5] ISO 3745[4.3] recommends that the microphone is traversed at constant speed using a turntable along multiple parallel circular paths regularly spaced on the test hemisphere (or sphere). Figure 4.2 gives the parallel circular paths for microphone traverses in a free field over a reflecting plane.

Example 4.1

Estimate the error in the sound-power level result in the 1250-Hz one-third octave band due to a limited number of measuring points for Ex. 3.1 at a confidence level of 97.5 per cent. Also, estimate the upper limit of the true sound-power level in the one-third octave band. The sound-pressure levels at the 12-point array are 76, 78, 79, 82, 83, 86, 88, 88, 89, 89, 90, and 92 dB.

Solution. Number of measuring points $n = 12$. In Table A.2.1 the one-tailed value of Student's t for $(12-1) = 11$ degrees of freedom which will be exceeded with the probability of $(100-97.5)$ per cent $=$ 2.5 per cent is found to be 2.201. From Ex. 3.1, the value of $p_{\mathrm{av}}^2/p_{\mathrm{ref}}^2$ is 5.31×10^8.

L_i (dB)	$\dfrac{p_i^2}{p_{ref}^2}=10^{L_i/10}$	$\dfrac{p_i^2}{p_{ref}^2}-\dfrac{p_{av}^2}{p_{ref}^2}$	$\left(\dfrac{p_i^2}{p_{ref}^2}-\dfrac{p_{av}^2}{p_{ref}^2}\right)^2$
76	3.98×10^7	-4.91×10^8	24.11×10^{16}
78	6.30×10^7	-4.68×10^8	21.90×10^{16}
79	7.94×10^7	-4.52×10^8	20.43×10^{16}
82	1.58×10^8	-3.73×10^8	13.91×10^{16}
83	1.99×10^8	-3.32×10^8	11.02×10^{16}
86	3.98×10^8	-1.33×10^8	1.77×10^{16}
88	6.30×10^8	$+0.99\times10^8$	0.98×10^{16}
88	6.30×10^8	$+0.99\times10^8$	0.98×10^{16}
89	7.94×10^8	$+2.63\times10^8$	6.92×10^{16}
89	7.94×10^8	$+2.63\times10^8$	6.92×10^{16}
90	1.00×10^9	$+4.69\times10^8$	22.00×10^{16}
92	1.58×10^9	$+10.49\times10^8$	1.10×10^{18}

$$\sum_{i=1}^{12} 10^{L_i/10}=6.37\times10^9 \qquad \sum_{i=1}^{12}\left(\frac{p_i^2}{p_{ref}^2}-\frac{p_{av}^2}{p_{ref}^2}\right)^2=2.41\times10^{18}$$

The standard deviation σ_E of the p_i^2/p_{ref}^2 values is given by

$$\sigma_E=\left[\frac{1}{n}\sum_{i=1}^{12}\left(\frac{p_i^2}{p_{ref}^2}-\frac{p_{av}^2}{p_{ref}^2}\right)^2\right]^{1/2}=[\tfrac{1}{12}(2.41\times10^{18})]^{1/2}=4.48\times10^8.$$

From eqn (4.2), the error in sound-power level at 97.5 per cent confidence level is

$$L_{W,true}-L_{W,measured}=10\log_{10}\left[1+n\sigma_E t\Big/\left(\sqrt{(n-1)}\sum_{i=1}^{n}10^{L_i/10}\right)\right]$$
$$=10\log_{10}\{1+12\times4.48\times10^8\times2.201/[\sqrt{(12-1)}\times6.37\times10^9]\}=1.93\text{ dB}.$$

Thus, noting that $L_{W,measured}=99.8$ dB from Ex. 3.1, the upper limit of the true sound power level at 97.5 per cent confidence level is

$$=L_{W,measured}+1.93=99.8+1.93=101.7\text{ dB}.$$

Example 4.2

Using Fig. 4.1, estimate the error in the sound-power level result for Ex. 3.3 due to the limited number of measuring points and estimate the upper limit of true sound-power level at 97.5 per cent confidence level. The sound-pressure levels in Ex. 3.3 are 60, 62, 64, 65, 66, 66, 68, 68, 69, 71, 72, and 72 dB.

Solution. Number of measuring points $n=12$. The spread of sound-pressure levels is $S=72-60=12$ dB. From Fig. 4.1 the error in the sound-power level for $n = 12$ and $S = 12$ is approximately equal (at 97.5 per cent confidence level) to 1.6 dB. From Ex. 3.3 the measured sound-power level $L_{W,\text{measured}} = 78.6$ dB. Thus, the upper limit of the true sound-power level at 97.5 per cent confidence level is

$$L_W = L_{W,\text{measured}} + 1.6 = 78.6 + 1.6 = 80.2 \text{ dB}.$$

4.2. Error due to 'far-field' assumption

All equations to calculate the sound-power output of a machine described in Chapter 3 are based on the 'far-field' assumption, by which we mean that (1) the instantaneous sound pressure and particle velocity at all measuring points are assumed to be in phase with each other; (2) the measuring surface is assumed to be at every point perpendicular to the direction of sound propagation; and (3) the sound intensity, which is the average rate of energy flow across a unit area, is assumed to be equal to $p^2/\rho c$. However, except for a progressive plane or spherical wave, there is in general a phase angle between the pressure and particle velocity, and the sound intensity is not exactly equal to $p^2/\rho c$. In fact, as will be discussed in Section 6.1, the sound intensity should be expressed as

$$I_{\mathbf{n}} = p v_{\mathbf{n}} \cos \phi \tag{4.5}$$

where p is the r.m.s. value of the sound pressure, $v_{\mathbf{n}}$ is the r.m.s. value of the particle velocity in the direction $\mathbf{n}$, which is perpendicular to the measuring surface concerned, and ϕ is the phase angle between the pressure and velocity. Therefore, there is an inherent error in the sound-power level result calculated from the 'far-field' assumption. This error can be expressed as

$$L_{W,\text{true}} - L_{W,\text{far-field}} = 10 \log_{10} \left(\frac{\sum_{i=1}^{n} p_i v_{\mathbf{n},i} \cos \phi \, dA_i}{\sum_{i=1}^{n} (p_i^2/\rho c) \, dA_i} \right) \text{dB}. \tag{4.6}$$

In practice, it is rather difficult to calculate the error given by eqn (4.6) accurately as

(1) The machine surface is quite often of irregular shape;

(2) The sound-pressure level measuring points, as determined by particular measurement arrays (such as were described in Chapter 3), do not in general bear the same relationship to the wave fronts of the propagating sound;

(3) The machine surface vibration modes may be very complicated and unknown;

(4) The available analytical expressions are derived for only a very few

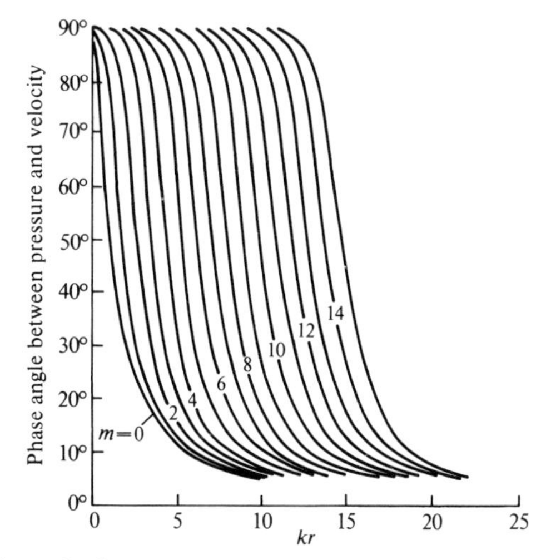

Fig. 4.3. Variation of phase angle between pressure and velocity with kr for various vibration modes.[4.6] $k=\omega/c$, $\omega=2\pi f$, c = speed of sound, r = radius, m = vibration mode number.

idealized shapes, e.g. a sphere and a cylinder, having certain specified modes of surface vibration.

Nevertheless, some results obtained from theoretical studies will be given in this section to show the basic factors involved in the inherent error due to the 'far-field' assumption.

Theoretical studies[4.6]–[4.8] on the radiation properties of spherical noise sources have shown that the phase angle between the pressure and velocity on a spherical surface enclosing a spherical noise source varies with the vibration mode m and the product of the radius of the measuring surface r and the wave number k ($k=\omega/c=2\pi f/c$). Figure 4.3 shows examples of the variation of the phase angle with kr and m.[4.6]

TABLE 4.2. Comparison of sound-power levels, L_W, calculated for different radii of circumscribing cylindrical surface for a small electric machine[4.9]

Frequency (Hz)	Based on $I=p^2/\rho c$			Based on $I=pv\cos\phi$			L_p, measured value* dB
	$r=1$ m	$r=0.5$ m	$r=0.25$ m	$r=1$ m	$r=0.5$ m	$r=0.25$ m	
550	33.6	34.1	36.3	30.0	29.7	30.7	32.0
1000	29.5	30.5	39.5	39.1	29.6	28.7	31.5
1500	27.7	27.1	27.9	26.9	26.6	26.7	27.9
2400	29.2	29.2	29.9	29.4	29.1	28.5	31.6

* Narrowband sound-power level using passband 6 per cent of centre frequency.

An investigation[4.9] on the sound radiation properties of a cylindrical surface enclosing a cylindrical electric machine found that the phase angle varied with the vibration mode, the location of the measuring point, the wave number, and the length and diameter of the cylindrical noise source. The analytical expressions for the sound pressure, particle velocity, and phase angle are given in ref. [4.9]. Table 4.2 shows the results of sound-power levels calculated from both the 'far-field' assumption and the accurate intensity formula as defined in eqn. (4.5) for a small machine. One sees that for a radius of 0.25 m at $f = 550$ Hz, which is not in the 'far field', the sound-power level based on $I = p^2/\rho c$ is much higher than the actual level.

Except for a few idealized cases, the evaluation of eqn (4.6) is extremely complicated since the three parameters, pressure, particle velocity, and phase angle, should be calculated at all measuring points. In order to estimate an approximate value for the inherent error in the sound power due to the 'far-field' assumption, it could be assumed that the sound intensity at all measuring points is equal to $(p_i^2/\rho c)$ times the cosine of the phase angle ϕ_{ref} at a key reference point on the measuring surface. Thus, eqn (4.6) becomes

$$L_{\text{W,true}} - L_{\text{W,far-field}} \approx 10 \log_{10}(\cos \phi_{\text{ref}}) \text{ dB}. \tag{4.7}$$

For short machines and small machines, the phase angle may be estimated from Fig. 4.3.

Figure 4.3 shows that the phase angle between the pressure and particle velocity varies from nearly 90° for very small kr values to less than 10° for large kr values. It is of interest to note that the phase angle is a function of the product of k and r, rather than of the individual values of either k or r. Other things being equal, if the frequency, hence the k value, is doubled, the same phase angle would be expected at a point which is at half the original radius, since the kr value at this new point is the same as at the original one.

For estimating the error in sound-power level due to the 'far-field' assumption, it is essential to check the kr value. For a vibration mode less than 4 and a kr value less than 5, Fig. 4.3 shows that the phase angle would be less than 35°, which corresponds approximately to a sound-power level error of less than 1 dB due to the 'far-field' assumption. On the other hand, if the kr value concerned is less than 1, a large error in the sound-power level due to the far-field assumption would be expected.

Example 4.3

A short electric machine has a predominant noise component at 100 Hz with a vibration mode of 2. Based on Fig. 4.3 estimate the phase angle between the sound pressure and particle velocity at a point R which is

1 m from the centre of the machine. Assuming that the sound intensity at all measuring points is equal to the product of $p_i^2/\rho c$ and the cosine of the phase angle at this point R, estimate the error in the sound-power level due to the 'far-field' assumption.

Solution. For the 100-Hz sound wave, the wave number is

$$k = \frac{\omega}{c} = \frac{2\pi f}{c} \approx \frac{2\pi(100)}{340} = 1.85.$$

At point R, the radius is 1.0 m and the value of $kr = 1.85$. From Fig. 4.3 the phase angle for $m = 2$ and $kr = 1.85$ is

$$\phi_R \approx 76°.$$

Assuming that the sound intensity at all measuring points is equal to $p_i^2 \cos \phi_R/\rho c$, the ratio of the true sound power to the sound power based on the 'far-field' assumption is

$$\frac{W_{\text{true}}}{W_{\text{far-field}}} = \frac{\sum p_i^2 \cos \phi_R/\rho c}{\sum p_i^2/\rho c} = \cos \phi_R = \cos 76° = 0.241.$$

Thus the approximate error in the sound-power level due to the 'far-field' assumption is

$$L_{W,\text{true}} - L_{W,\text{far-field}} = 10 \log_{10} \frac{W_{\text{true}}}{W_{\text{far-field}}}$$

$$= 10 \log_{10}(\cos \phi_R) = 10 \log_{10}(0.241) = -6.2 \text{ dB}.$$

4.3. Error due to ground reflection

When the source is placed on a hard surface such as concrete, the noise measured is the sum of the direct noise and the reflected noise. A hard surface reflects incident sound waves without a change of phase of the pressure wave. It may therefore be replaced by an image source emitting sound with the same magnitude and phase as the real source at a distance below the boundary ground plane equal to the distance of the real source above the plane (Fig. 4.4).

If the distance between the two sources is small compared with the distance to the point at which the sound pressure is to be measured, the difference between the lengths of paths from the two sources is approximately $2h \sin \theta$. The phase angle difference between the direct and reflected wave is $4\pi h \sin \theta/\lambda$ and, ignoring the difference in inverse-square-law attenuation for the two waves, the total pressure at the

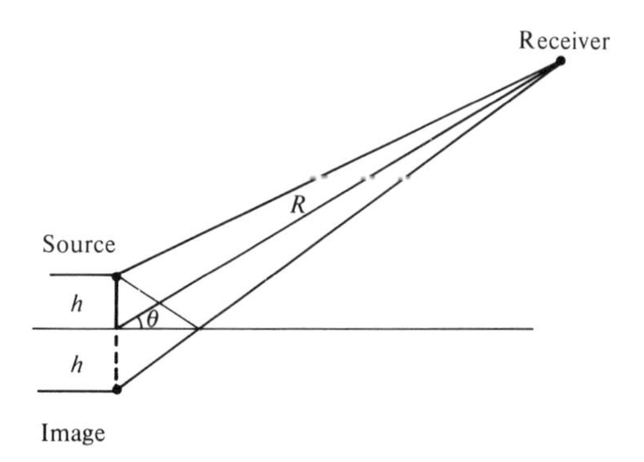

Fig. 4.4. Geometry of sound source, image, and receiver.

measuring point is given by

$$p_{\mathrm{h}}(t)=\sqrt{2}p_{\mathrm{s}}\left[\cos \omega t+\cos\left(\omega t+\frac{4\pi h \sin \theta}{\lambda}\right)\right]$$
$$=2\sqrt{2}p_{\mathrm{s}} \cos \frac{2\pi h \sin \theta}{\lambda} \cos \left(\omega t+\frac{2\pi h \sin \theta}{\lambda}\right). \tag{4.8}$$

The r.m.s. pressure at a point on a hemispherical surface with its centre midway between its source and its image is therefore given by $2p_{\mathrm{s}} |\cos (2\pi h \sin \theta/\lambda)|$, where p_{s} is the pressure which would be obtained at the same distance if the source were radiating spherically. The level of the mean-square sound pressure obtained from the pressures at n points on a hemispherical surface will therefore be

$$\bar{L}_{\mathrm{p,h}}=\bar{L}_{\mathrm{p,s}}+\log_{10}\left[\frac{4}{n}\sum_{i=1}^{n} \cos^2\left(\frac{2\pi h \sin \theta_i}{\lambda}\right)\right] \tag{4.9}$$

where $\bar{L}_{\mathrm{p,s}}$ is equal to $10 \log_{10}[(p_{\mathrm{s}}^2)_{\mathrm{av}}/p_{\mathrm{ref}}^2]$. The mean sound-pressure level $\bar{L}_{\mathrm{p,s}}$ corresponding to the power radiated spherically by the source is obtained from

$$L_{\mathrm{W,s}}=\bar{L}_{\mathrm{p,s}}+10 \log_{10}4\pi r^2. \tag{4.10}$$

Similarly, the sound-power level obtained from measurements on a hemispherical surface over the ground plane is

$$L_{\mathrm{W,h}}=\bar{L}_{\mathrm{p,s}}+10 \log_{10}\left[\frac{4}{n}\sum_{i=1}^{n} \cos^2\left(\frac{2\pi h \sin \theta_i}{\lambda}\right)\right]+10 \log_{10}2\pi r^2. \tag{4.11}$$

Therefore the power level measured by this hemispherical method over the hard ground surface differs from the true power level which would

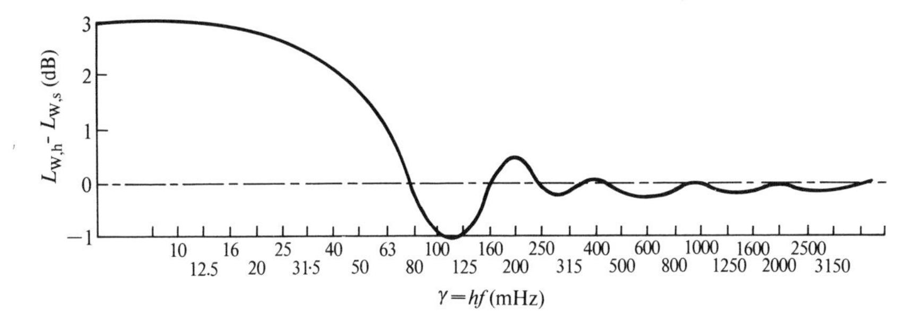

Fig. 4.5. Error in hemispherical sound-power level.

have been obtained by spherical measurements in a free field by

$$L_{W,h} - L_{W,s} = 10 \log_{10}\left[\frac{2}{n}\sum_{i=1}^{n}\cos^2\left(\frac{2\pi h \sin\theta_i}{\lambda}\right)\right]. \tag{4.12}$$

If the number of measuring points is very large, this difference reduces to

$$L_{W,h} - L_{W,s} = 10 \log_{10}\left[1 + \frac{c}{4\pi\gamma}\sin\left(\frac{4\pi\gamma}{c}\right)\right]. \tag{4.13}$$

The above equation is plotted in Fig. 4.5, where $\gamma = hf$ mHz.

When the ground surface is soft and porous, e.g. sand and grassy turf, the ground reflection is incomplete.[4.10] Since there are very few data on the properties of such ground, accurate correction for reflection from soft and porous ground is difficult. However, interesting information is available for sound-power measurements made on flat grass ground using hemispherical and parallelepiped measuring surfaces. Hübner and Meurers[4.11] showed that, for a reference sound source, the sound-power level determined in a semi-anechoic chamber with a 1-meter distance parallelepiped measuring surface did not differ by more than 0.4 dB from the sound-power level determined from measurements on flat grassy ground using the same parallelepiped surface. However, considerable errors in sound-power levels were introduced by the outdoor measurements on flat grassy ground when using greater measurement distances, e.g. 4 m or 8 m. In order to reduce the inaccuracy in sound-power level caused by the effect of natural ground reflection, a sound absorbing pad can be placed on the ground near the test machine so that the relevant ground is made non-reflecting. Moore[4.12] used a polyurethane foam pad to reduce the effect of ground reflection for outdoor fan noise testing and found that the absorbing pad reduced the effect of ground reflection to approximately ±1 dB for the frequency range from 100 Hz to 20 kHz.

Example 4.4

Using an eight-point 1 m hemispherical array enclosing a machine on a hard ground in an anechoic chamber, the sound-pressure levels obtained in the one-third octave 50-Hz band were 78, 76, 68, 70, 72, 73, 76, and 74 dB at points 1, 2, 3, 4, 5, 6, 7, and 8 respectively. The locations of these points are given in the following table. The centre of the machine is 0.15 m above the ground and the machine base is placed at the centre of the sphere. Estimate the error in the sound-power level due to ground reflection and calculate the sound-power level in the one-third octave band.

8-point array table

Point	x/r	y/r	z/r
1	0.97	0	0.25
2	0	−0.97	0.25
3	−0.97	0	0.25
4	0	+0.97	0.25
5	0.63	0	0.78
6	0	−0.63	0.78
7	−0.63	0	0.78
8	0	+0.63	0.78

Solution. The wavelength $\lambda = c/f = 340/50 = 6.8$ m. At point 1, from the above table, $\theta = \arctan(0.25/0.97) = 14°30'$; therefore,

$$\cos^2\left(\frac{2\pi h \sin\theta}{\lambda}\right) = \cos^2\left(\frac{2\pi(0.15)\sin(14°30')}{6.8}\right)$$
$$= 0.999.$$

Since points 2, 3, and 4 have the same θ value as point 1, their $\cos^2(2\pi h \sin\theta_i/\lambda)$ values are all equal to 0.999. At point 5, from the above table, $\theta = \arctan(0.78/0.63) = 51°06'$; therefore,

$$\cos^2\left(\frac{2\pi h \sin\theta_5}{\lambda}\right) = \cos^2\left(\frac{2\pi(0.15)\sin(51°06')}{6.8}\right) = 0.988.$$

Since points 6, 7, and 8 have the same θ as point 5, their $\cos^2(2\pi h \sin\theta_i/\lambda)$ values are all equal to 0.988. Thus,

$$\sum_{i=1}^{8} \cos^2\left(\frac{2\pi h \sin\theta_i}{\lambda}\right) = 4(0.999 + 0.988) = 7.948.$$

From eqn (4.12), the error due to ground reflection is

$$L_{W,h}-L_{W,s}=10\log_{10}\left[\frac{2}{n}\sum_{i=1}^{n}\cos^2\left(\frac{2\pi h\sin\theta_i}{\lambda}\right)\right]$$
$$=10\log_{10}\{\tfrac{2}{18}\times 7.948\}\approx 3\text{ dB}.$$

The mean sound-pressure level on the hemispherical surface is, from eqn (1.10)

$$\bar{L}_{p,h}=10\log_{10}\left(\frac{1}{8}\sum_{i=1}^{8}10^{0.1L_{p,i}}\right)$$
$$=10\log_{10}[\tfrac{1}{8}(10^{7.8}+10^{7.6}+10^{6.8}+10^{7.0}+10^{7.2}+10^{7.3}+10^{7.6}+10^{7.4})]$$
$$=74.4\text{ dB}.$$

The sound-power level is, from eqn (4.12) and taking into account the error due to ground reflection,

$$L_W=L_{W,s}=L_{W,h}-3=\bar{L}_{p,h}+10\log_{10}2\pi r^2-3$$
$$=74.4+10\log_{10}[2\pi(1.0)^2]-3=79.4\text{ dB}.$$

4.4. Other problems

When using sound measuring equipment it is always necessary to check its calibration. Barometric pressure, temperature, and humidity may affect the calibration of the system and appropriate adjustments should be made immediately before the use of the equipment as instructed in the equipment manufacturer's manual. Furthermore, attention should be paid to the equipment electrical floor noise if low noise levels are to be measured and to wind noise for outdoor measurements. For noise having considerable high-frequency components, microphone alignment is significant. These points will now be discussed further.

4.4.1. *Electrical noise floor*

Care should be taken in using sound measuring equipment for measuring low noise levels. The electrical noise floor of the instrumentation should be checked, since it sets the lower limit to the level of the signals that can be analysed with accuracy. In checking the noise floor of a complete sound measuring system, the microphone should be shielded from the acoustic background noise. Some acoustic calibrators give good acoustic isolation at high frequencies. A typical example of electrical noise floor for a sound-level meter is shown in Fig. 4.6.[4.13] Figure 4.6 also shows the acoustic background noise levels in the space and the electrical noise floor for the sound-level meter with the microphone replaced by an equivalent electrical impedance.

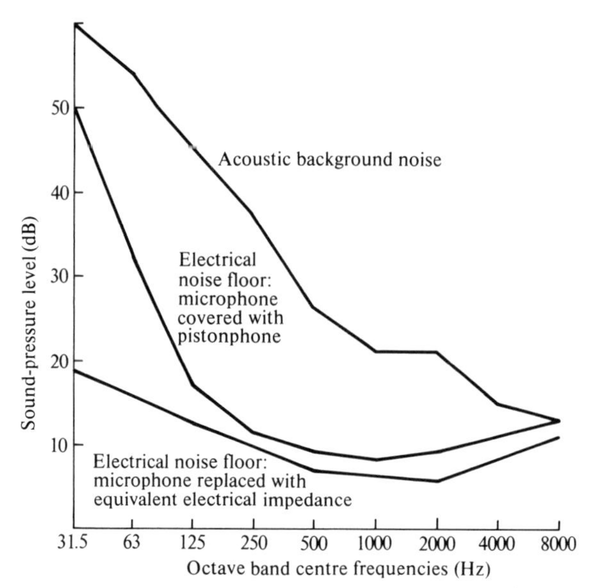

Fig. 4.6. Example of electrical noise floor (Reproduced by permission of IEEE, Copyright © 1970 IEEE).[4.13]

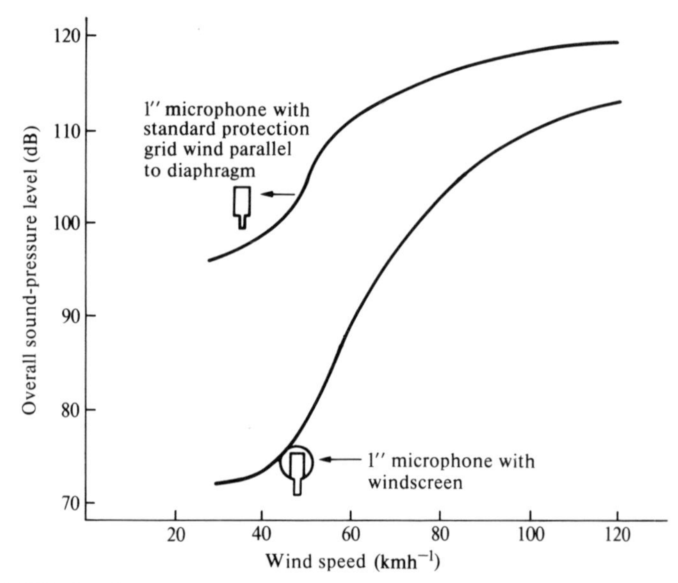

Fig. 4.7. Wind noise as a function of wind speed (Reproduced by permission of Brüel & Kjaer).[4.14]

4.4.2. *Wind noise*

For outdoor measurements consideration should be given to wind noise. The wind noise level increases with wind speed as shown in Fig. 4.7,[4.14] which also shows the recorded noise level when a spherical foam material windscreen of about 11.5 cm in diameter is attached to an ordinary microphone. When the windscreen is used, the air flow turbulence is located farther away from the microphone and the noise is thus attenuated when it reaches the diaphragm of the microphone. Analyses of wind noise in one-third octave bands are shown in Fig. 4.8[4.14] for wind speeds of 30 km h^{-1} and 120 km h^{-1} with and without the windscreen.

It is always advisable, under outdoor windy conditions, for two sets of noise measurements to be made, one with the machine under test and one without the machine, using a given microphone–windscreen measuring system. If the differences in the two sets of readings are less than 3 dB, no meaningful noise measurement results can be expected. Otherwise Fig. 3.17 can be used to make the necessary corrections.

4.4.3. *Microphone alignment*

The effect of microphone alignment on noise measurement results is usually negligible. However, the free-field response of a 2.54 cm (1 inch) condenser microphone may have an error of about 1.7 dB at 3000 Hz and

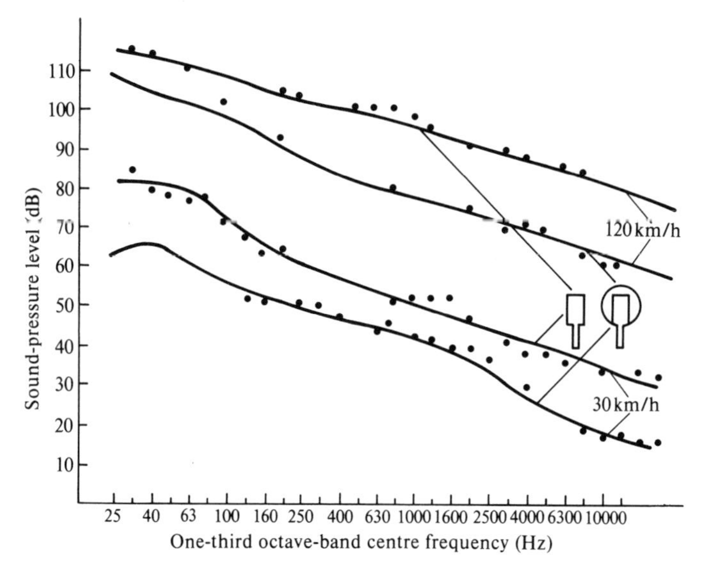

Fig. 4.8. Frequency spectra of wind noise, measured in one-third octave bands, with wind direction parallel to the diaphragm of the microphone (Reproduced by permission of Brüel & Kjaer).[4.14]

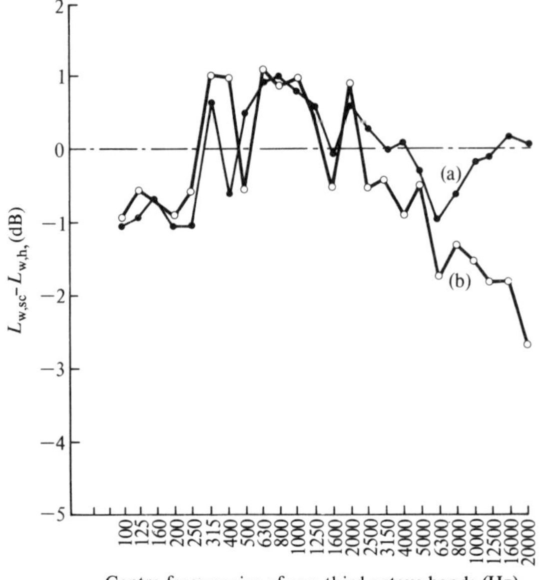

Fig. 4.9. Difference in sound-power level in one-third octave bands from semicylindrical measurement in anechoic chamber $L_{w,sc}$ and from hemispherical measurement in anechoic chamber $L_{w,h}$.[4.2] (a) Microphone pointing to centre of machine; (b) microphone pointing along line perpendicular to measuring surface.

an error of about 2.9 dB at 4000 Hz when the angle of incidence of the noise is 90° with respect to the normal to the diaphragm of the microphone. Thus, at frequencies higher than 3000 Hz, a considerable error in the sound power results may be introduced if the microphone is not pointing towards the centre of the noise source, i.e. roughly in the direction of the sound propagation.

An investigation of this was made on a small electric machine placed in an anechoic chamber.[4.2] Figure 4.9 gives the sound-power levels in one-third octave bands for two cases: (1) the microphone pointing to the centre of the machine; (2) the microphone pointing along a line perpendicular to the measuring surface. In both cases the measuring surface was the same semi-cylindrical surface, specified by the BEAMA document.[4.15] The results show that for frequencies up to 6300 Hz there was less than 1 dB difference in the sound-power levels for the two cases. This might be due to the averaging effect taken over the different measurement points. However, measurements made with the microphone pointing along a line perpendicular to the measuring surface gave results which were much too low for frequencies higher than 10 kHz. It appears necessary, therefore, that the microphone should point roughly towards the centre of the noise source, especially for sources which have considerable high frequency components.

4.4.4. *Ambient temperature and pressure*

According to ISO 3745[4.3] the correction in sound-power results for ambient temperature and pressure is given by

$$\Delta = 10 \log_{10}\left[\frac{B}{1000}\sqrt{\left(\frac{293}{273+\theta}\right)}\right]$$

where Δ is the correction in dB to be added to the sound-power level, B is the barometric pressure in millibars, and θ is the temperature in degrees Celsius.

References

[4.1] Weatherburn, C. E. (1962). Mathematical statistics. Cambridge University Press, Cambridge.

[4.2] Ellison, A. J., Moore, C. J., and Yang, S. J. (1969). Methods of measurement of acoustic noise radiated by an electric machine. *Proc. I.E.E.* **116,** 1419–31.

[4.3] ISO (1977). *Determination of sound power levels of noise sources—precision methods for anechoic and semi-anechoic rooms*, ISO 3745. ISO, Geneva.

[4.4] Hübner, G., Schmidt, H., Herbert, M., Meurers, H., and Woehle, K. K. (1976). Investigations of the establishment of International Standards for the measurement of noise emitted by machines—newer aspects of uncertainties in the determination of the sound power. *Proceedings of Inter-Noise 76*, pp. 405–10. Institute of Noise Control Engineering, New York.

[4.5] Moore, C. J. and Ellison, A. J. (1968). An acoustic power measuring device. *J. sci. Instrum.* (*J. Phys. E*), Series 2, **1,** 659–61.

[4.6] Ellison, A. J. and Moore, C. J. (1969). Calculation of acoustic power radiated by short electric machines. *Acustica* **21,** 10–15.

[4.7] Hübner, G. (1973). Analysis of errors in measuring machine noise under free field conditions. *J. acoust. Soc. Am.* **54,** 967–77.

[4,8] Hoffman, R., Jordan, H., and Weis, M. (1966). Ersatzstrahler zur Ermittlung der Schalleistung von rotierenden elektrischen Machinen. *Larmbekampfung* **1,** 7–11.

[4.9] Ellison, A. J. and Yang, S. J. (1971). Calculation of acoustic power radiated by an electric machine. *Acustica* **25,** 28–34.

[4.10] Onclay, P. B. (1970). Propagation of jet engine noise near a porous surface. *J. sound Vib.* **13,** 27–35.

[4.11] Hübner, G. and Meurers, H. (1975). Investigations of sound propagation over non-ideal reflecting planes and the influence of the measurement distance on the accuracy of sound power determination of sound sources operating outdoors. *Congress Proceedings of Le Bruit des Machines et l'Environment*, pp. 559–67. Federation of Acoustical Societies of Europe, Paris.

[4.12] Moore, C. J. (1971). A solution to the problem of measuring the sound field of a source in the presence of a ground surface. *J. Sound Vib.* **16**(2), 269–82.

[4.13] Bruce, R. D. (1970). Measurement of noise. *IEEE Trans. Geoscience Electronics* **GE-8**, 130–8.
[4.14] Skode, F. (1966). Windscreening of outdoor microphones. *B & K Tech. Rev.* **1,** 3–9.
[4.15] BEAMA (1967). *BEAMA recommendations for the measurement and classification of acoustic noise from rotating electric machines*, BEAMA publication 225. BEAMA, London.

Further reading for Chapter 4

[1] Concawe (1979). *Method for determining the sound power levels of flares used in refineries, chemical plants and oilfields*, Concawe report no. 2/79. Concawe, The Hague.
[2] Fukuhara, H. and Ohkuma, T. (1981). Influence of wind on the measurement of infrasound. *Proceedings of Inter-Noise 81*, pp. 955–8. Nederlands Akoestisch Genootschap, Delft.
[3] Nemerlin, J. (1981). 'In-situ' power measurement and ground effect—Part 1. *Proceedings of Inter-Noise 81*, pp. 929–34. Nederlands Akoestisch Genootschap, Delft.
[4] Mertens, C., Soubrier, D., and Henderiechx, F. (1981). 'In-situ' power measurement and ground effect—Part 2: Testing and accuracy of the method. *Proceedings of Inter-Noise 81*, pp. 935–40. Nederlands Akoestisch Genootschap, Delft.
[5] Tichy, J. (1981). Precision of the sound power determination in a reverberant room with rotating diffuser. *Proceedings of Inter-Noise 81*, pp. 925–8. Nederlands Akoestisch Genootschap, Delft.

5. Measurements on nominally identical small machines

5.1. Statement of problem

A considerable variation in noise data between the members of a group of nominally identical small machines, e.g. small electric machines, business machines, and domestic appliances, has been observed. The difference in sound-pressure level data taken at the same points for nominally identical small machines may well exceed 20 dB (see Fig. 5.1). The problem thus arises of how to make a judgement of the noise characteristics of mass-produced small machines. To test every machine or appliance in a production line is too time-consuming. It therefore appears more practicable to tackle the problems under the following headings by statistical techniques.

1. The presentation of the noise data for 'identical' machines referred to a large population;
2. The degree of added precision gained by increasing the number of machines tested and the number of measuring points around each machine;
3. The extent to which the test data will be affected by the inability to control and duplicate test conditions. (Examples of this are errors introduced by variations in the measuring equipment and the ambient conditions and the taking of readings by different people.)

In this chapter the special considerations involved in the presentation of noise data for small nominally identical machines are discussed.

The sound-pressure levels produced by small machines or appliances are sometimes not large enough to be measurable in an ordinary semi-reverberant room as the background noise level, especially that at lower frequencies, may be as high as that radiated by the machines. Figure 5.2 shows that the background noise in the laboratory used is much higher than the noise from a small machine at frequencies lower than about 3 kHz. It is necessary, therefore, except for high-speed small machines, to use either an anechoic chamber or a reverberant room.

The parameters describing the noise characteristics of the machines as a whole are conveniently the mean sound level (A), and the mean sound-power levels in various frequency bands, both referred to the large population. The mean sound-power level can be found from eqn (1.9). However, $\bar{L}_p$ is now the level of the mean-square sound pressure in a frequency band referred to the whole measuring surface and to the whole

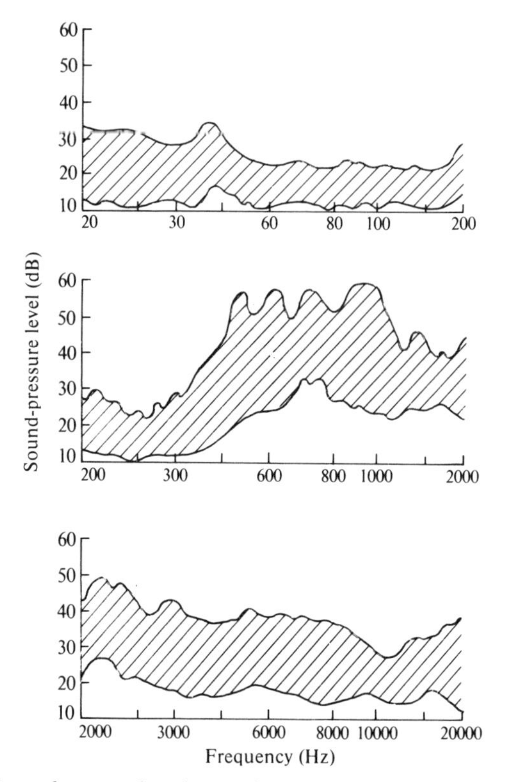

Fig. 5.1. Variation of narrowband sound pressure levels between 10 nominally identical motors having ball bearings at same point in acoustic field (passband 6 per cent of centre frequency).[5.1]

population of machines. If the 'A' weighting network is used, the parameter is the mean sound level (A) referred to the whole population of machines. As recommended by BEAMA,[5.2] this mean sound level (A) can be expressed at a reference radius of 3 m as

$$L_{A,3m} = \bar{L}_A + 20 \log_{10} \frac{r}{3} \tag{5.1}$$

where r is the equivalent radius in metres of the measuring surface.

Such mean levels can be obtained only when a large number of motors are tested each at a very large number of points, and this is impracticable. Nevertheless, the upper and lower limits of the mean sound level (A) and the mean sound-power levels can be estimated at given confidence levels, as described in Sections 5.2 and 5.3, from a relatively small number of measurements.

For convenience, the following discussion will be based on the

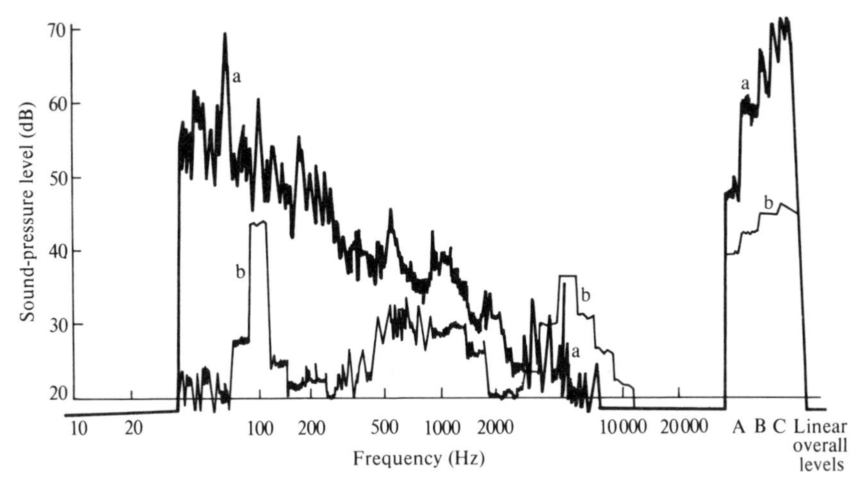

Fig. 5.2. Background noise and noise emitted from a small machine.[5.1] (a) Background noise in laboratory in one-third octave bands; (b) noise from 0.25-kW motor in one-third octave bands at 30 cm from its centre, measured in anechoic chamber.

measured noise data of two groups of nominally identical fractional-horse-power machines. Ten nominally identical fractional-horse-power motors having sleeve-bearings and 10 similar motors having ball-bearings (otherwise the same as the first batch) were chosen randomly from the production line in a factory where these machines were mass produced. Sound-pressure level measurements were made for all machines on a spherical measuring surface in a free-field environment.

Figure 5.1 shows the spread in s.p.l. from 20 Hz to 20 kHz between 10 nominally identical ball-bearing machines at the same point in the noise field using a filter having a pass bandwidth of 6 per cent of the centre frequency. It is seen that the spread in s.p.l. at several frequencies for these machines well exceeds 20 dB. The mean sound level A referred to 3 m produced by each of ten nominally identical machines is given for both types of machine in Table 5.1. The values given here are the upper

TABLE 5.1. Variation of mean sound level (A) (dB(A)) referred to 3 m radius between nominally identical machines (at 97.5 per cent confidence level)[5.1]

Type of bearing	No. of motor									
	1	2	3	4	5	6	7	8	9	10
Sleeve	26.1	29.9	27.9	24.2	31.3	28.7	24.6	27.8	28.7	29.9
Ball	31.8	37.5	34.2	35.1	32.4	37.1	40.4	31.5	35.9	36.8

TABLE 5.2. Variation of sound level (A) (dB(A)) at various points on the measuring surface (sleeve-bearing machines)[5.1]

	Point*					
Number of motor	Top point	Right (facing shaft) point	Non-shaft end point	Bottom point	Left (facing shaft) point	Shaft end point
1	45.5	43	39	39.5	43	43
2	46	48	38	42.5	47.5	42.5
3	44.5	47.5	40	41	45.5	40.5
4	42.5	43	40	40	42.5	40.5
5	46.5	51	42.5	42.5	49	45
6	46.5	46	43.5	43.5	46	40.5
7	42.5	42.5	41	42.5	42.5	40.5
8	43	47.5	40	39.5	45	41.5
9	48.5	48.5	41.5	43	46.5	41.5
10	48	47	42.5	46	47.5	41

* Each point is at 38.2 cm from the centre of the machine.

limits of the mean sound levels, each calculated from the six readings and the maximum error expected at 97.5 per cent confidence level, as described in Section 4.1. Table 5.2 shows the variations in value of sound level A between six different points (top, bottom, front, back, and two sides) at the same distance from the centre of the sleeve-bearing machines. Table 5.3 shows the variations with time, over a period of about 15 seconds, of the sound-pressure level at specified points around a sleeve-bearing machine. In addition to the variations with the machine, with position, and with time, there was quite often a fluctuation of a few

TABLE 5.3. Variation with time over a period of 15 seconds of sound level (A) (dB(A)) at specified points for one sleeve-bearing machine[5.1]

	Point					
Observation during 15 s	Top point	Right (facing shaft) point	Non-shaft end point	Bottom point	Left (facing shaft) point	Shaft end point
No. 1	46	47.5	38	42.5	47.5	43
No. 2	46	48	38	42.5	47	43
No. 3	45.5	47	38.5	42.5	47.5	43
No. 4	46	48	38.5	42.5	47.5	42.5
No. 5	46.5	48.5	38	42.5	48	42.5

TABLE 5.4. Variation in sound level (A) (dB(A)) taken by different observers from the same pen recordings[5.1]

Observer	No. of motor										Maximum deviation over 10 motors
	1	2	3	4	5	6	7	8	9	10	
1	47	53	50	48	47	64	58	48	55	51	17
2	47	53	51	48	47	64	56	47	55	51	17
3	47	53	50	48	46.5	64	56	47.5	54	51	17.5

dB or more seen at a particular point on the recording paper at a particular time and the reading taken from the recording paper varied with the observer as shown in Table 5.4. This makes clear the importance of a statistical expression for the noise produced by nominally identical machines, which could not be accepted or rejected by a customer from a manufacturer on the basis of a single noise measurement. Specifications in contracts for relatively large numbers of such small machines should perhaps be written in statistical form.

5.2. The mathematical model

As small machines are mass produced, the limited number of machines randomly chosen for noise measurement should be regarded as a sample from a large population. If J identical machines are each tested at I points on a measuring surface and K observations are made at each point for each machine, the mathematical model representing the ijkth reading, y_{ijk}, is assumed to be

$$y_{ijk} = \bar{\mu} + a_i + b_j + c_{ij} + e_{ijk}, \tag{5.2}$$

where $\bar{\mu}$ = overall mean of noise readings for a large population, a_i = deviation from mean introduced by differences between readings at different points in the field of the same machine, b_j = deviation from mean introduced by differences between readings at the same point for different machines, c_{ij} = deviation from mean introduced by effect of interaction between machines and points, and e_{ijk} = the random error peculiar to the observation (personal bias, variation in ambient conditions, and slight changes in conditions in measuring equipment).

All quantities in eqn (5.2) must be expressed in terms of the square of the sound pressure, instead of the sound-pressure level, as the square of the sound pressure (or the ratio of the square of the sound pressure to the square of the reference sound pressure) is proportional to sound intensity if measurements are made in the far field, and these quantities can be added together to give the total sound intensity.

In order to assess the relative importance of various factors in the noise measurements, the analysis of variance[5.2,5.3] is used. From the analysis of variance and Snedecor's F test (applied here to Tables 5.3 and 5.4), it is found that the factors for personal bias on reading from the pen recordings and for differences between observations at different times (after the initial 20-min running period) are not significant, but the factors allowing for differences between individual machines and for differences between readings taken at individual points on the measuring surface are significant.

Table 5.5 gives the results of the analysis of variance for sound-level A data taken from 10 sleeve-bearing machines at six points each and for two observations at different instants. In Table 5.5 the variance ratio F values in the sixth column are obtained by dividing the mean-squares in the fourth column by the mean squares of the residual error because it is necessary to determine whether the variations due to points, motors, and their interaction are small and can therefore be included in the residual error. However, tables of the F-distribution show these F values to be higher than the corresponding upper 1 per cent points and thus the factors allowing for the variations between motors and points and their interaction are significant and must be separately assessed. Table 5.6 gives the similar results of the analysis of variance for sound level (A) data taken from 10 ball-bearing machines at six points, each with two observations at different instants. The assumed mathematical model is therefore an appropriate one for these machines. Similar tables of

TABLE 5.5. Analysis of variance for sleeve-bearing machines, based on sound level (A) (dB(A)). All values refer to p^2/p_{ref}^2 (10 motors, 6 points, 2 observations)

Source of variation	Sum of squares SS	Degrees of freedom ν	Mean square SS/ν	Expected value of mean square $E(MS)$	Ratio F MS/MS_e	Upper 1 per cent point of F-distribution
Among points (Factor A)	2.426×10^{10}	5	4.835×10^{9}	$\sigma_e^2+2\sigma_{AB}^2+20\sigma_A^2$	222.1	3.34
Among motors (Factor B)	1.781×10^{10}	9	1.979×10^{9}	$\sigma_e^2+2\sigma_{AB}^2+12\sigma_B^2$	90.1	2.72
Interaction of A with B	2.055×10^{10}	45	4.567×10^{8}	$\sigma_e^2+2\sigma_{AB}^2$	20.8	1.89
Residual error	1.312×10^{9}	60	2.186×10^{7}	σ_e^2	1.0	

TABLE 5.6. Analysis of variance for ball-bearing machines, based on sound level (A) (dB(A)). All values refer to p^2/p^2_{ref}, (10 motors, 6 points, 2 observations)

Source of variation	Sum of squares SS	Degrees of freedom ν	Mean square SS/ν	Expected value of mean square $E(MS)$	Ratio F MS/MS_e	Upper 1 per cent point of F-distribution
Among points (factor A)	2.296×10^{11}	5	4.592×10^{10}	$\sigma_e^2+2\sigma_{AB}^2+20\sigma_A^2$	15.1	3.34
Among motors (factor B)	1.610×10^{12}	9	1.789×10^{11}	$\sigma_e^2+2\sigma_{AB}^2+12\sigma_B^2$	58.9	2.72
Interaction of A with B	1.338×10^{12}	45	2.974×10^{10}	$\sigma_e^2+2\sigma_{AB}^2$	9.8	1.89
Residual error	1.821×10^{11}	60	3.035×10^{9}	σ_e^2	1.0	

analysis of variance can be made for noise data obtained from other nominally identical machines and appliances.

5.3. The confidence limit of the mean sound-power level

Our aim here is to find the confidence limits for the mean-square sound pressure, hence the confidence limits for the mean sound-power level or mean sound level, at a given confidence level. From eqn (5.2), it can be shown (see Appendix 3) that the standard deviation of the total error in p^2_{av}/p^2_{ref} is given by

$$\sigma_{tot}=\left(\frac{\sigma_A^2}{I}+\frac{\sigma_B^2}{J}+\frac{\sigma_{AB}^2}{IJ}+\frac{\sigma_e^2}{IJK}\right)^{1/2} \tag{5.3}$$

where σ_A^2 is the variance for the effect of points (factor A), σ_B^2 the variance for the effect of machines (factor B), σ_{AB}^2 the variance for the interaction of points and machines, and σ_e^2 the variance for residual errors.

According to Bulmer[5.5] and Scheffe,[5.6] the upper and lower limits for the component variances σ_A^2, σ_B^2, and σ_{AB}^2 can be found from the values of mean squares, degrees of freedom as given in the table of analysis of variance (e.g. Table 5.5), and the F-distribution as follows.

If two mean squares MS_1 and MS_2 are independently distributed with ν_1 and ν_2 degrees of freedom, respectively, and the expected value of the mean square MS_1 is the sum of the expected value of the mean square MS_2 and $C\sigma_1^2$ (i.e. $E(MS_1)=E(MS_2)+C\sigma_1^2$), the upper limit for variance σ_1^2 with an approximate confidence level of $100(1-\alpha)$ per cent is

$$\sigma_{1,u}^2 = \frac{MS_2}{C}\left[F_{\alpha;\infty,\nu_1}\frac{MS_1}{MS_2} - 1 + \frac{MS_2}{F_{\alpha;\nu_2,\nu_1}MS_1}\left(1 - \frac{F_{\alpha;\infty,\nu_1}}{F_{\alpha;\nu_2,\nu_1}}\right)\right] \quad (5.4)$$

for $MS_1/MS_2 > 1/F_{\alpha;\nu_2,\nu_1}$ and is equal to zero for $MS_1/MS_2 < 1/F_{\alpha;\nu_2,\nu_1}$.

The lower confidence limit for σ_1^2, with approximate confidence level of $100(1-\alpha)$ per cent is

$$\sigma_{1,l}^2 = \frac{MS_2}{C}\left[\frac{MS_1}{F_{\alpha;\nu_1,\infty}MS_2} - 1 + \frac{F_{\alpha;\nu_1,\nu_2}MS_2}{MS_1}\left(\frac{F_{\alpha;\nu_1,\nu_2}}{F_{\alpha;\nu_1,\infty}} - 1\right)\right] \quad (5.5)$$

for $MS_1/MS_2 > F_{\alpha;\nu_1,\nu_2}$ and is equal to zero for $MS_1/MS_2 < F_{\alpha;\nu_1,\nu_2}$.

In eqn (5.4), $F_{\alpha;\infty,\nu_1}$ is the upper α point of the F-distribution with ∞ and ν_1 degrees of freedom and $F_{\alpha;\nu_2,\nu_1}$ is the upper point of the F-distribution with ν_2 and ν_1 degrees of freedom. In eqn (5.5), $F_{\alpha;\nu_1,\infty}$ is the upper α point of the F-distribution with ν_1 and ∞ degrees of freedom and $F_{\alpha;\nu_1,\nu_2}$ is the upper point of the F-distribution with ν_1 and ν_2 degrees of freedom. For calculating the upper limit of σ_A^2, MS_1 is replaced by MS_A and MS_2 by MS_{AB}. For calculating the upper limit of σ_B^2, MS_1 is replaced by MS_B and MS_2 by MS_{AB}. For calculating the upper limit of σ_{AB}^2, MS_1 is replaced by MS_{AB} and MS_2 by MS_e (see Ex. 5.1).

Assuming the distribution of the residual errors to follow the normal (Gaussian) distribution, the upper and lower limits for σ_e^2 can be found by using the χ^2-distribution table at a given confidence level.

Using the upper limits of these variances, the upper limit of the standard deviation of the total error in the value of p_{av}^2/p_{ref}^2, $\sigma_{tot,u}$, can be obtained from eqn (5.3). By the multiplication rule of probabilities,[5.4] the confidence level for $\sigma_{tot,u}$ is equal to the product of the confidence levels of the four individual component variances. In other words, the confidence level of $\sigma_{tot,u}$ is equal to $(1-\alpha_A)(1-\alpha_B)(1-\alpha_{AB})(1-\alpha_e)$, where $(1-\alpha_A)$ is the confidence level for the upper limit of σ_A, $\sigma_{A,u}$, $(1-\sigma_B)$ the confidence level for $\sigma_{B,u}^2$, $(1-\alpha_{AB})$ the confidence level for $\sigma_{AB,u}^2$, and $(1-\alpha_e)$ the confidence level for $\sigma_{e,u}^2$.

From the central limit theorem,[5.4] the distribution of the total error in the measured mean-square sound pressure can be regarded as Gaussian with zero mean. Using the well known characteristics of the Gaussian distribution and the multiplication rule of probabilities, the total error in the measured mean-square sound pressure will be less than $\gamma\sigma_{tot,u}$, with a confidence level of $P(-\infty \leqslant z \leqslant \gamma)(1-\alpha_A)(1-\alpha_B)(1-\alpha_{AB})(1-\alpha_e)$, where $P(-\infty \leqslant z \leqslant \gamma)$ is the probability that the standard normal variable z is less than γ (γ is any positive value). The upper limit of the level of the mean-square sound pressure in a frequency band or in A-weighting is thus given by

$$\bar{L}_{\mathrm{p,u}} = 10 \log_{10}\left(\frac{(p_i^2)_{\mathrm{av,measured}}}{p_{\mathrm{ref}}^2} + \gamma\sigma_{\mathrm{tot,u}}\right) \tag{5.6}$$

or

$$\bar{L}_{\mathrm{A,u}} = 10 \log_{10}\left(\frac{(p_{i,\mathrm{A}}^2)_{\mathrm{av,measured}}}{p_{\mathrm{ref}}^2} + \gamma\sigma_{\mathrm{tot,u}}\right) \tag{5.7}$$

with a confidence level of $P(-\infty \leqslant z \leqslant \gamma)(1-\alpha_{\mathrm{A}})(1-\alpha_{\mathrm{B}})(1-\alpha_{\mathrm{AB}})(1-\alpha_{\mathrm{e}})$. Then the upper limit for the mean sound-power level in a frequency band or the mean sound level (A) referred to 3 m is readily obtained from eqns (1.9) or (5.1) (see Ex. 5.2). The lower limits can be found similarly, but usually only the upper limits are of interest.

For the sleeve-bearing machines, from measurements with a total sample size of 120 (i.e. $N = IJK = 6\,\mathrm{pts} \times 10\ \mathrm{motors} \times 2$ readings each), the upper limit for the mean sound level (A) referred to a 3-m reference radius was found to be 30.0 dB(A) with a 90 per cent confidence level and 31.6 dB(A) with a 97.5 per cent confidence level. The upper limits for the sound-power levels in various frequency bands can be found by the same technique. After the sound-power level in each octave band of the audible spectrum is found for the whole population, the corresponding noise rating for the population may be derived in the same manner as that used for a single machine.

One is now in a position not only to compare these upper limits with the upper limits of the sound level in A-weighting referred to a reference radius, or sound-power level in a frequency band (as probably to be specified by various national and international standards), but also to estimate the percentage of the machines in a population which gives a noise higher than a specified value. For example, for the sleeve-bearing machines about 10 per cent give a mean sound level A at 3 m greater than 30.0 dB(A) and about 2.5 per cent of them give a mean sound level A greater than 31.6 dB(A). If any alteration is made in design or manufacturing procedures on the production line, it is thus possible to find statistically the effect on the noise characteristics of the machines by the above analysis of two sets of measurements, one taken before and one after the change.

Direct verification of the above method needs many measurements and is practically difficult. However, by comparing the above upper limits with those in Table 5.1, it was found that none of the 10 sleeve-bearing machines produced a mean sound level exceeding the upper limit at the 97.5 per cent confidence level while only one of them gave a mean level exceeding the upper limit at the 90 per cent confidence level.

The upper limits of mean sound level (A) and components of variance obtained by this method for different numbers of measurements

TABLE 5.7. Variation of upper limits of mean sound level (A) (dB(A)) and components of variance with sample size for sleeve-bearing machines

Upper limits of $\bar{L}_A$ or components of variance	Sample size N $(I\times J\times K)$						Confidence level (per cent)
	60 (6×5×2)*	96 (6×8×2)	120 (6×10×2)	180 (6×10×3)	240 (6×10×4)	300 (6×10×5)	
$\bar{L}_{A,3m,u}$	33.3	31.5	31.6	31.6	31.6	31.6	97.5
$\sigma^2_{A,u}$	3.78×10^9	2.64×10^9	2.93×10^9	2.90×10^9	2.92×10^9	2.94×10^9	99.5
$\sigma^2_{B,u}$	5.02×10^9	1.19×10^9	8.20×10^8	7.79×10^8	7.99×10^8	7.86×10^8	99.5
$\sigma^2_{AB,u}$	9.16×10^8	4.81×10^8	4.13×10^8	3.67×10^8	3.59×10^8	3.66×10^8	99.5

* Number of points × number of motors × number of readings each.

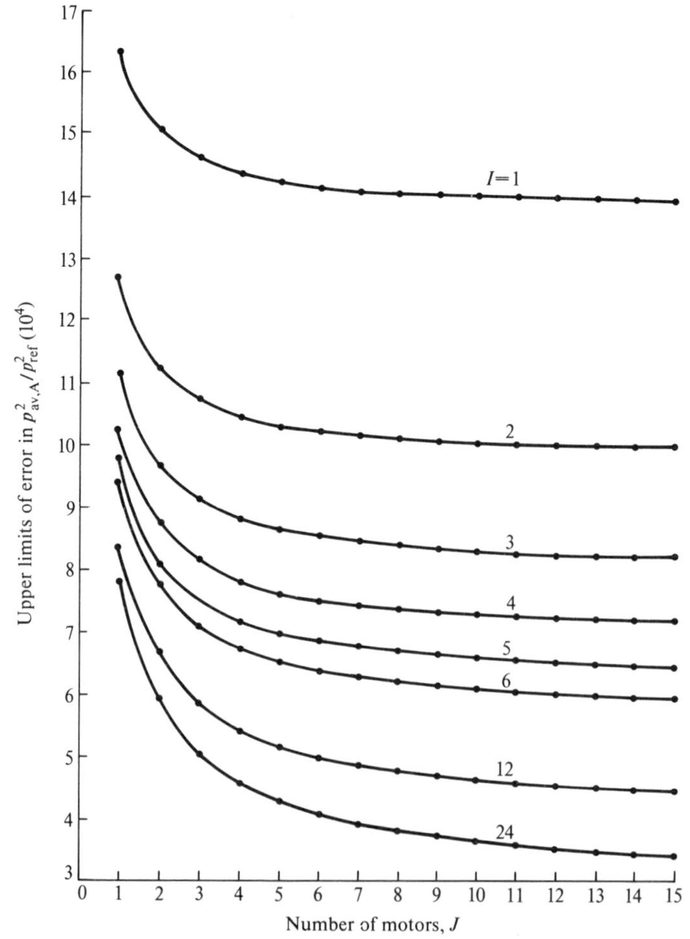

Fig. 5.3. Variation of upper limits of error in $p^2_{av,A}/p^2_{ref}$ with number of motors J and number of points I for sleeve-bearing machines (two observations at each point) at 97.5 per cent confidence level.[5.1]

on sleeve-bearing machines are given in Table 5.7, which shows practically consistent results for sample sizes equal to or greater than 120. Therefore the confidence limits for the components of variance obtained from a set of measurements with a sufficiently large sample size, say a minimum of 120, can be used to estimate the total error involved in other designs of measurement procedure.

Figures 5.3 and 5.4 show the examples of the upper limits of the error in $p^2_{av,A}/p^2_{ref}$ for the sleeve-bearing and ball-bearing machines under test with a confidence level of 97.5 per cent for different numbers of points and numbers of motors tested. From these curves it is possible to

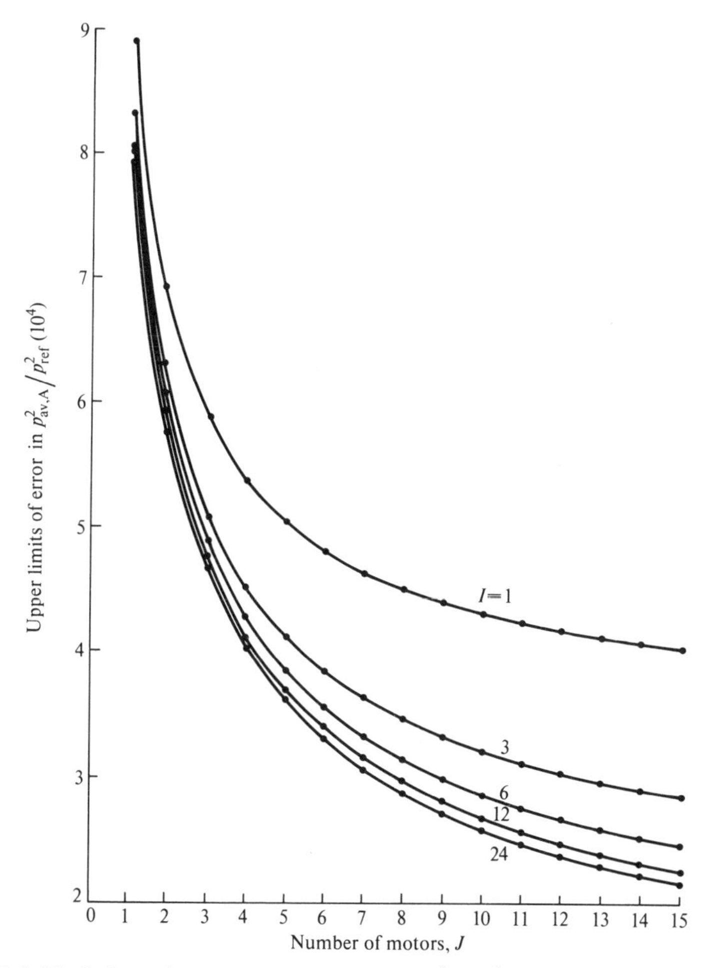

Fig. 5.4. Variation of upper limit of error in $p^2_{av,A}/p^2_{ref}$ with number of motors J and number of points I for ball-bearing machines (two observations at each point) at 97.5 per cent confidence level.[5.1]

estimate the necessary number of motors to be tested, and the number of points at which measurements must be taken, to achieve a desired degree of accuracy.

It is seen that, for the sleeve-bearing machines, increasing the number of measurement points can increase the degree of accuracy of the measurements more effectively than can increasing the number of motors. For the ball-bearing machines, an increase in the number of motors is more beneficial. The reason is the greater variability of ball bearings as compared with sleeve-bearings.

The optimal design of measurement procedure is found by minimizing the standard deviation of the total error for a fixed total number of measurements. If the product of number of points I, number of motors J, and number of observations K, is a constant N, the suitable choice between number of points and number of motors for a fixed number of observations is obtained by differentiating eqn (5.3) with respect to number of points. Thus the number of points is given by

$$I = \frac{N}{K} \frac{\sigma_{A,u}}{\sigma_{B,u}} \tag{5.8}$$

where $\sigma_{A,u}$ and $\sigma_{B,u}$ are the upper limits of σ_A and σ_B, respectively at the same confidence level.

5.4. Approximate calculating procedures

In order to simplify the above calculating procedure, approximations have been studied. Instead of using only values of p_i^2/p_{ref}^2 in the calculating procedures, data in s.p.l. values were used directly to calculate confidence limits for components of variance and the standard deviation of the total error. The upper limit of the level of mean-square sound pressure, or the mean sound level (A), with a given confidence level, was found by adding the corresponding upper limit of the total error to the measured arithmetic mean level. The results obtained by these approximate procedures for the sleeve-bearing machines are compared in Fig. 5.5 with those obtained by the previous method for different sample sizes of measurements (number of motors$>$5, number of observations$>$2, number of points$=$6). It is seen that, for sample sizes greater than 120, the approximate procedures give results about 3 dB higher than those obtained by the previous method; for the smaller sample sizes the approximate procedures give much higher upper limits.

Example 5.1

The analysis of variance table for the noise measurement data of a group of identical small machines is given in Table 5.5. Calculate (1)

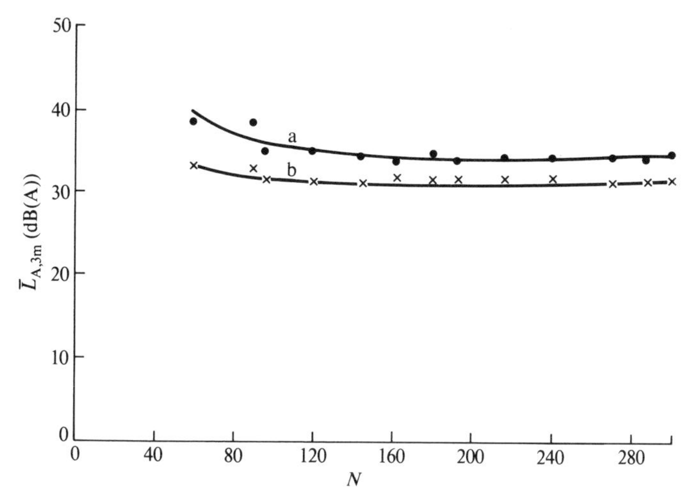

Fig. 5.5. Variation of upper limits of mean sound level $\bar{L}_{A,3m}$ with sample size N for sleeve bearing machines at 97.5 per cent confidence level.[5.1] (a) By the approximate method of Section 5.4; (b) by the more accurate method of Section 5.3.

the upper limit for the variance of the effect of points $\sigma^2_{A,u}$; (2) the upper limit for the variance of the effect of machines $\sigma^2_{B,u}$; and (3) the upper limit for the variance of point–machine interaction $\sigma^2_{AB,u}$, each with an approximate confidence level of 99 per cent.

Solution. (1) To find the upper limit for σ^2_A at 99 per cent confidence level.

$$100(1-\alpha) \text{ per cent} = 99 \text{ per cent}$$

so

$$\alpha = 0.01.$$

From Table 5.5,

$$MS_1 = MS_A = 4.835\times10^9, \qquad \nu_1 = 5;$$
$$MS_2 = MS_{AB} = 4.567\times10^8, \qquad \nu_2 = 45;$$
$$E(MS_A) = E(MS_{AB}) + 20\sigma^2_A, \quad \text{and}$$
$$C = 20.$$

From the F-distribution table[5.7–5.9]

$$F_{\alpha;\infty,\nu_1} = F_{0.01;\infty,5} = 9.02$$

and

$$F_{\alpha;\nu_2,\nu_1} = F_{0.01;45,5} = 9.27.$$

From eqn (5.4), at 99 per cent confidence level, the upper limit for σ^2_A

is

$$\sigma^2_{A,u} = \frac{MS_{AB}}{C}\left[F_{0.01;\infty,5}\left(\frac{MS_A}{MS_{AB}}\right) - 1 + \frac{MS_{AB}}{F_{0.01;45,5}MS_A}\left(1 - \frac{F_{0.01;\infty,5}}{F_{0.01;45,5}}\right)\right]$$
$$= \frac{4.567\times 10^8}{20}\left[9.02\frac{4.853\times 10^9}{4.567\times 10^8} - 1 + \frac{4.567\times 10^8}{9.27(4.853\times 10^9)}\left(1 - \frac{9.02}{9.27}\right)\right]$$
$$= 2.17\times 10^9.$$

(2) To find the upper limit for σ^2_B with 99 per cent confidence level. From Table 5.5,

$$MS_1 = MS_B = 1.979\times 10^9, \qquad \nu_1 = 9;$$
$$MS_2 = MS_{AB} = 4.567\times 10^8, \qquad \nu_2 = 45;$$
$$E(MS_B) = E(MS_{AB}) + 12\sigma^2_B, \quad \text{and}$$
$$C = 12.$$

From the F-distribution table,

$$F_{\alpha;\infty,\nu_1} = F_{0.01;\infty,9} = 4.31$$

and

$$F_{\alpha;\nu_2,\nu_1} = F_{0.01;45,9} = 4.54.$$

From eqn (5.4), the upper limit for σ^2_B is

$$\sigma^2_{B,u} = \frac{MS_{AB}}{C}\left[F_{0.01;\infty,9}\left(\frac{MS_B}{MS_{AB}}\right) - 1 + \frac{MS_{AB}}{F_{0.01;45.9}MS_B}\left(1 - \frac{F_{0.01;\infty,9}}{F_{0.01;45.9}}\right)\right]$$
$$= \frac{4.567\times 10^8}{12}\left[4.31\frac{1.979\times 10^9}{4.567\times 10^8} - 1 + \frac{4.567\times 10^8}{4.54\times 1.979\times 10^9}\left(1 - \frac{4.31}{4.54}\right)\right]$$
$$= 6.73\times 10^8.$$

(3) To find the upper limit for σ^2_{AB} with 99 per cent confidence level. From Table 5.5,

$$MS_1 = MS_{AB} = 4.567\times 10^8, \qquad \nu_1 = 45;$$
$$MS_2 = MS_e = 2.186\times 10^7, \qquad \nu_2 = 60;$$
$$E(MS_{AB}) = E(MS_e) + 2\sigma^2_{AB}, \qquad \text{and}$$
$$C = 2$$

From the F-distribution table,

$$F_{\alpha;\infty,\nu_1} = F_{0.01;\infty,45} = 1.75$$

and

$$F_{\alpha;\nu_2,\nu_1} = F_{0.01;60,45} = 1.97.$$

From eqn (5.4), at the 99 per cent confidence level, the upper limit of σ^2_{AB} is

$$\sigma^2_{\mathrm{AB,u}} = \frac{MS_{\mathrm{e}}}{C}\left\{F_{0.01;\infty,45}\left(\frac{MS_{\mathrm{AB}}}{MS_{\mathrm{e}}}\right) - 1 + \frac{MS_{\mathrm{e}}}{F_{0.01;60,45}MS_{\mathrm{AB}}}\left(1 - \frac{F_{0.01;\infty,45}}{F_{0.01;60,45}}\right)\right\}$$

$$= \frac{2.186\times 10^7}{2}\left\{1.75\frac{4.567\times 10^8}{2.186\times 10^7} - 1 + \frac{2.186\times 10^7}{1.97(4.567\times 10^8)}\left(1 - \frac{1.75}{1.97}\right)\right\}$$

$$= 3.88\times 10^8.$$

Example 5.2

For a group of nominally identical machines, the table for analysis of variance is given in Table 5.5. Using the results in Ex. 5.1 and given that the measured $(p^2_{\mathrm{i,A}})_{\mathrm{av}}/p^2_{\mathrm{ref}}$ at a radius of $r = 0.38$ m is 2.86×10^4, calculate (1) the upper limit for this standard deviation of the total error in $(p^2_{\mathrm{i,A}})_{\mathrm{av}}/p^2_{\mathrm{ref}}$, i.e. $\sigma_{\mathrm{tot,u}}$, at the 96 per cent confidence level; (2) the upper limit of $(p^2_{\mathrm{i,A}})_{\mathrm{av}}/p^2_{\mathrm{ref}}$ at the 95 per cent confidence level; and (3) the upper mean sound level (A) referred to 3 m with an approximate confidence level of 95 per cent.

Solution. (1) From Ex. 5.1, the upper limits for σ^2_{A}, σ^2_{B}, and σ^2_{AB} are

$$\sigma^2_{\mathrm{A,u}} = 2.17\times 10^9 \text{ at } 1-\alpha_{\mathrm{A}} = 99 \text{ per cent confidence level,}$$

$$\sigma^2_{\mathrm{B,u}} = 6.73\times 10^8 \text{ at } 1-\alpha_{\mathrm{B}} = 99 \text{ per cent confidence level,}$$

and

$$\sigma^2_{\mathrm{AB,u}} = 3.88\times 10^8 \text{ at } 1-\alpha_{\mathrm{AB}} = 99 \text{ per cent confidence level.}$$

The confidence level for $\sigma_{\mathrm{tot,u}}$ is required to be 96 per cent, so

$$(1-\alpha_{\mathrm{A}})(1-\alpha_{\mathrm{B}})(1-\alpha_{\mathrm{AB}})(1-\alpha_{\mathrm{e}}) = 0.96$$

and

$$1-\alpha_{\mathrm{e}} = \frac{0.96}{(0.99)^3} \approx 0.99, \qquad \text{i.e. } \alpha_{\mathrm{e}} = 0.01.$$

From the χ^2 Table[5.7–5.9] for 60 degrees of freedom,

$$\chi^2_{(1-0.01)} = \chi^2_{(0.99)} = 37.485.$$

From Table 5.5 the unbiased sample variance for residual error is $\sigma_e = 2.186 \times 10^7$. Thus the upper limit for population residual error variance is

$$(\sigma^2_{e,u})_{popul.} = \frac{(60)(2.186 \times 10^7)}{37.485}$$
$$= 3.5 \times 10^7 \text{ at } (1-\alpha_e) = 99 \text{ per cent confidence level.}$$

From eqn (5.3), the upper limit of the standard deviation of the total error in $(p^2_{i,A})_{av}/p^2_{ref}$ is, at a confidence level of 96 per cent,

$$\sigma_{tot,u} = \left(\frac{\sigma^2_{A,u}}{I} + \frac{\sigma^2_{B,u}}{J} + \frac{\sigma^2_{AB,u}}{IJ} + \frac{\sigma^2_{e,u}}{IJK}\right)^{1/2}$$
$$= \left(\frac{2.17 \times 10^9}{6} + \frac{6.73 \times 10^8}{10} + \frac{3.88 \times 10^8}{60} + \frac{3.5 \times 10^7}{120}\right)^{1/2}$$
$$= 2.09 \times 10^4.$$

(2) Let $P(-\infty \leqslant z \leqslant \gamma)(1-\alpha_A)(1-\alpha_B)(1-\alpha_{AB})(1-\alpha_e) = 0.95$.

$$P(-\infty \leqslant z \leqslant \gamma) \approx 0.99$$

where $P(-\infty \leqslant z \leqslant \gamma)$ is the probability that the standard normal variable z is less than γ. From tables of the standard normal distribution,[5.7]–[5.9] the value of γ to give $P(-\infty \leqslant z \leqslant \gamma) = 0.99$ is found to be 2.33.

Thus, the upper limit of the total error in $(p^2_{i,A})_{av}/p^2_{ref}$ is, at an approximate confidence level of 95 per cent,

$$\gamma\sigma_{tot,u} = 2.33 \times 2.09 \times 10^4 = 4.87 \times 10^4.$$

The upper limit of $(p^2_{i,A})_{av}/p^2_{ref}$ is, at the 95 per cent confidence level,

$$\frac{(p^2_{i,A})_{av,u}}{p^2_{ref}} = \frac{(p^2_{i,A})_{av,measured}}{p^2_{ref}} + \gamma\sigma_{tot,u}$$
$$= 2.86 \times 10^4 + 4.87 \times 10^4$$
$$= 7.73 \times 10^4.$$

(3) We are 95 per cent confident that the upper limit of the mean sound level (A) at radius $r = 0.38$ m is

$$\bar{L}_{A,u} = 10 \log_{10}\left(\frac{(p^2_{i,A})_{av,u}}{p^2_{ref}}\right)$$
$$= 10 \log_{10}(7.73 \times 10^4) = 48.9 \text{ dB(A)}.$$

The upper limit for the mean sound level (A) referred to 3 m radius is, at the 95 per cent confidence level,

$$\bar{L}_{\mathrm{A,3m,u}} = \bar{L}_{\mathrm{A,u}} + 20\log_{10}\frac{r}{3}$$

$$= 48.9 + 20\log_{10}\frac{0.38}{3} \approx 31\ \mathrm{dB(A)}.$$

For comparison, the measured mean sound level (A) referred to 3 m radius is given below

$$\bar{L}_{\mathrm{A,3m,measured}} = \bar{L}_{\mathrm{A,measured}} + 20\log_{10}\frac{r}{3}$$

$$= 10\log_{10}\left(\frac{(p^2_{\mathrm{i,A}})_{\mathrm{av,measured}}}{p^2_{\mathrm{ref}}}\right) + 20\log_{10}\frac{r}{3}$$

$$= 10\log_{10}(2.86\times10^4) + 20\log_{10}\frac{0.38}{3}$$

$$= 26.7\ \mathrm{dB(A)}.$$

Example 5.3

Using the residual error data given in Table 5.5, determine the 90 per cent confidence interval for the population residual error variance.

Solution. From Table 5.5, equating the expected value of the mean square for residual errors with the mean square value for residual errors, we get the unbiased sample residual error variance

$$\sigma^2_{\mathrm{e}} = 2.186\times10^7$$

and the degrees of freedom, $\nu = 60$.

$$100(1-2\alpha) = 90 \text{ per cent}, \quad \text{so} \quad \alpha = 0.05.$$

From tables of the χ^2 distribution,[5.7]–[5.9] for 60 degrees of freedom,

$$\chi^2_{(\alpha)} = \chi^2_{(0.05)} = 79.082$$

and

$$\chi^2_{(1-\alpha)} = \chi^2_{(0.95)} = 43.188.$$

Thus we are 90 per cent confident that the population residual error variance is within the limits

$$\frac{(60)(2.186\times10^7)}{79.082} < (\sigma_{\mathrm{e}})_{\mathrm{popul.}} < \frac{(60)(2.186\times10^7)}{43.188},$$

i.e. $1.66\times10^7 < (\sigma_{\mathrm{e}})_{\mathrm{popul.}} < 3.04\times10^7$.

References

[5.1] Ellison, A. J. and Yang, S. J. (1970). Acoustic noise measurements on nominally identical small electrical machines. *Proc. IEE* **117,** 555–60.
[5.2] BEAMA (1967). *BEAMA recommendations for the measurement and classification of acoustic noise from rotating electrical machines,* BEAMA publication 225. BEAMA, London.
[5.3] Dixon, W. J. and Massey, F. J. (1951). *Introduction to statistical analysis.* McGraw-Hill, New York.
[5.4] Richards von Mises (1964). *Mathematical theory of probability and statistics.* Academic Press, London.
[5.5] Bulmer, M. G. (1957). Approximate confidence limits for components of variance. *Biometrika* **44,** 159–67.
[5.6] Scheffe, H. (1959). *The analysis of variance.* John Wiley, New York.
[5.7] Brownlee, K. A. (1965). *Statistical theory and methodology in science and engineering.* John Wiley, New York.
[5.8] Hughes, A. and Grawoig, D. (1971). *Statistics—a foundation for analysis.* Addison-Wesley, Reading, Massachusetts.
[5.9] Beyer, W. H. (1966). *Handbook of tables for probability and statistics.* Chemical Rubber Co., Cleveland, Ohio.

6. Sound-intensity measurement

6.1. Sound intensity

In Section 1.4 we considered the calculation of the sound-power level emitted from a machine from the mean sound-pressure level measured over a surface enclosing that machine (see eqn (1.9)). However, in deriving eqn (1.9) one has to make the following assumptions:

(a) The measuring surface enclosing the machine is at every point perpendicular to the direction of sound propagation;
(b) The sound pressure and particle velocity are in phase with one another at every point and the particle velocity is equal to $p/\rho c$.

In practice, most machines have irregular shapes and various parts of a machine may emit noise in different directions. Thus the exact directions of the sound propagated around a machine are usually unknown. Furthermore, as described in Section 4.2, the sound pressure and particle velocity are, in general, not in phase with one another, especially for low frequencies and in the vicinity of the machine. To avoid the errors caused by these factors, it is necessary to use sound-intensity measurement instead of sound-pressure measurement.

Sound intensity, defined as the average rate of sound energy across a unit area, can be expressed as

$$I_{\mathbf{r}} = \frac{1}{T}\int_0^T p(t)v_{\mathbf{r}}(t)\,\mathrm{d}t \quad \mathrm{W\,m^{-2}}, \tag{6.1}$$

where $I_{\mathbf{r}}$ is the sound intensity in any specified direction $\mathbf{r}$ at a point in the sound field, $p(t)$ is the instantaneous sound pressure at the point $v_{\mathbf{r}}(t)$ is the component of instantaneous particle velocity in the direction $\mathbf{r}$ and T is the measurement period. The sound intensity represents the sound power transmitted in a specified direction through a unit area normal to this direction. Based on this definition of sound intensity, the sound power output of a machine can be expressed as

$$P = \int_A I_{\mathbf{n}}\,\mathrm{d}A \quad \mathrm{W} \tag{6.2}$$

where A is the total area of any surface enclosing the machine and $I_{\mathbf{n}}$ is the sound intensity in a direction $\mathbf{n}$ normal to an area element $\mathrm{d}A$. It should be emphasized that the sound intensity $I_{\mathbf{n}}$ is a vector quantity. Figure 6.1 shows a general case in which the real direction of sound propagation $\mathbf{R}$ is different from the direction $\mathbf{n}$ normal to the area

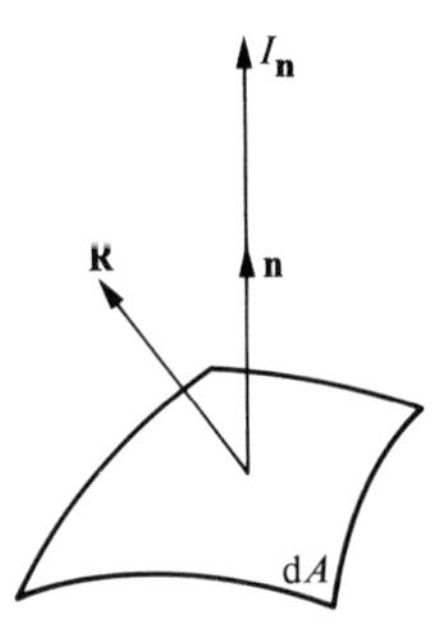

Fig. 6.1. Sound intensity in direction normal to the surface: **n**—the direction normal to area element dA; **R**—the real direction of sound propagation.

element dA. The actual sound power passing through dA is simply $I_{\mathbf{n}}\,\mathrm{d}A$, irrespective of the real direction of sound propagation.

If the surface enclosing a machine is divided into n equal and small parts and $I_{\mathbf{n}}$ is measured at each part, the total power emitted by the machine can be found from

$$P \approx \sum_{i=1}^{n} I_{\mathbf{n},i}\,\mathrm{d}A. \tag{6.3}$$

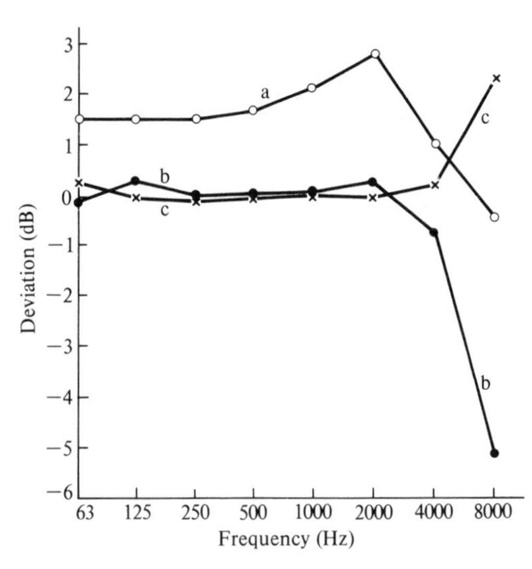

Fig. 6.2. Effect of shape of measurement surface and of background noise on determination of sound power. (Deviations from free-field measurements over a spherical surface, $A = A_{\mathrm{N}}$). (a) Sound-pressure measurements over a tetrahedral surface, $A \neq A_{\mathrm{N}}$; (b) sound-intensity measurements over a tetrahedral surface; (c) sound-intensity measurements over a spherical surface in the presence of background noise.[6.1]

The accuracy in sound power obtained from eqn (6.3) is not affected by the shape of the measuring surface since $I_{\mathbf{n},i}$, the intensity normal to the chosen measuring surface, is used at each point directly. However, the sound power obtained from eqn (1.4) is subject to the error introduced by the measuring surfaces not being perpendicular to the direction at each point.

Tests have been made[6.1] to compare the effect of the shape of the measuring surface on the sound power result. A small omnidirectional sound source radiating spherically was tested in an anechoic chamber. Both sound-intensity measurements and sound-pressure measurements were made on two measuring surfaces, one spherical surface concentric to the source and the other a tetrahedral surface enclosing the source. The difference in the sound-power results obtained from the sound-intensity measurements using these two measuring surfaces was negligibly small up to 4 kHz (see Fig. 6.2); the much larger error at 8 kHz was introduced by the fact that the polar response for oblique incidence of both pressure and velocity microphones deviated considerably from the ideal.[6.1] However, considerable errors were found in the sound-power results obtained from the sound-pressure measurements using the tetrahedral surface.

6.2. Sound-intensity measuring methods

Sound-intensity measuring equipment should sense both the sound pressure and particle velocity and transform them into two corresponding signals. Three different measuring methods will be briefly discussed.

6.2.1. *Pressure and velocity direct measurement method*

Figure 6.3 shows the schematic diagram of a type of sound-intensity measuring equipment based on the direct measurement of pressure and velocity in the noise field. The pressure and velocity signals are fed into a multiplier and its output is integrated and averaged.

In Fig. 6.3, the preamplifiers, bandpass filters, multiplier, integrator,

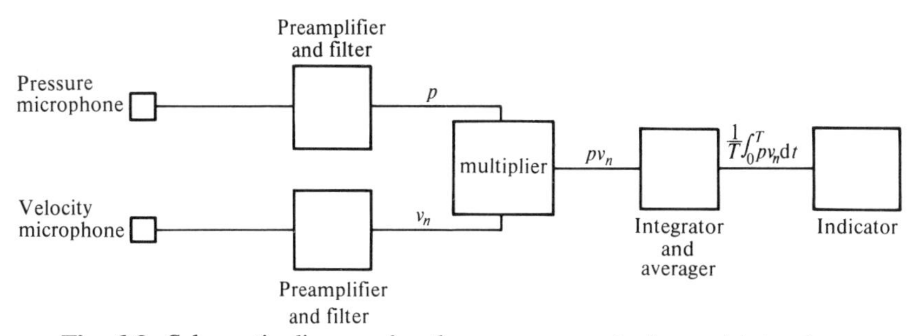

Fig. 6.3. Schematic diagram for the measurement of sound intensity.

and indicator are standard solid-state devices for performing these functions. For the intensity equipment developed by van Zyl and Anderson,[6.1] the pressure sensor was a 7-mm diameter electret microphone (see Section 7.1.2) and the velocity sensor a ribbon velocity microphone. The diaphragm of the electret microphone was facing the ribbon of the velocity microphone with a spacing of 2 mm. This intensity-measuring equipment had an accuracy, including the microphones, of ±1 dB over the frequency range of 50 Hz to 4 kHz.

The major drawback of a ribbon velocity microphone is that it is too delicate for use out of doors. The ribbon is fragile, very sensitive to wind and shock, and this microphone tends to produce a highly distorted signal in the presence of even the slightest wind.

There are other types of equipment for particle-velocity measurement, e.g. the hot-wire velocity sensor[6.2] and the pressure-gradient type.[6.3]–[6.6]

The hot-wire sensor has not become a commercial device. However, owing to the advances in digital signal-processing techniques, the two-microphone pressure-gradient velocity measurement has been successfully incorporated in commercial sound-intensity measuring systems.

6.2.2. *Two-pressure microphone method*

The principle of the two-pressure microphone method is briefly described as follows. The particle velocity in a sourceless region in a given direction $\mathbf{r}$ can be expressed as

$$V_{\mathbf{r}}(t) = -\frac{1}{\rho}\int \frac{\partial p}{\partial r}\,\mathrm{d}t$$
$$\approx -\frac{1}{\rho}\int \frac{p_2(t)-p_1(t)}{\Delta \mathbf{r}}\,\mathrm{d}t \tag{6.4}$$

where ρ is the density of air and $p_1(t)$ and $p_2(t)$ are the instantaneous pressure values measured by two small omnidirectional pressure microphones at two points separated by a distance of $\Delta\mathbf{r}$ (see Fig. 6.4). Equation (6.4) gives an approximate expression for the particle velocity in the direction $\mathbf{r}$ at a point midway between the two pressure measuring points. The pressure at the midpoint between the two points is approximated by

$$p(t) \approx \frac{p_1(t)+p_2(t)}{2}. \tag{6.5}$$

The sound intensity value in the direction $\mathbf{r}$ at the midpoint between the two microphones can be found by the integration of the product of $p(t)$ and $V_{\mathbf{r}}(t)$, based on eqns (6.4), (6.5), and (6.1). This integration can

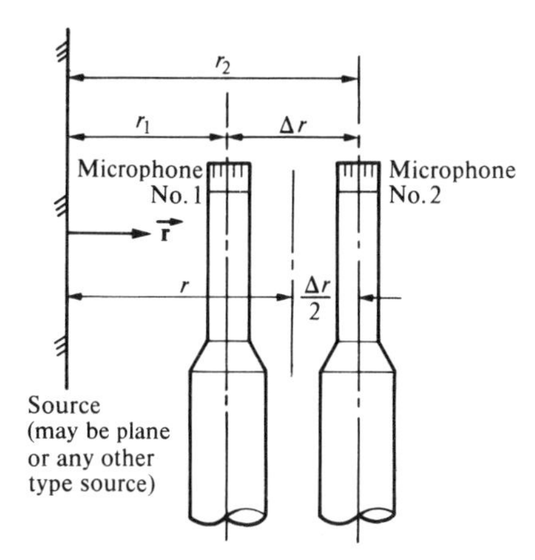

Fig. 6.4. Locations of two pressure microphones with respect to the source. r_1 = distance from source to microphone 1; r_2 = distance from source to microphone 2; Δr = separation between two microphones; $r = (r_1 + r_2)/2$.

readily be carried out by a two-channel spectrum analyser (see Section 7.5) or a commercial sound-intensity measuring system, using digitized $p_1(t)$ and $p_2(t)$ values.

The use of a two-channel spectrum analyser for determining sound intensity is based on Fahy's development of an intensity formulation in terms of cross-spectral densities.[6.8] Fahy has shown that the sound intensity is related to the cross-spectral density of $p_1(t)$ and $p_2(t)$ by the expression

$$I \approx \frac{1}{2\pi\rho\,\Delta \mathbf{r}} \int_0^\infty \left(\frac{Q(p_1 p_2)}{f}\right) \mathrm{d}f \tag{6.6}$$

where $Q(p_1 p_2)$ is the imaginary part of the cross-spectral density of $p_1(t)$ and $p_2(t)$ at the two microphone positions.

The two-microphone sound-intensity method has an inherent error caused by the finite approximation in eqns (6.4) and (6.5). Thompson and Tree[6.9] studied this error for three ideal sound sources. For a point monopole source, the finite-difference approximation error is found to be

$$\begin{aligned} L_e &= 10 \log_{10}\left(\frac{I_{app}}{I_{ex}}\right) \\ &= 10 \log_{10}\left[\left(\frac{\sin(k\,\Delta r)}{k\,\Delta r}\right)\left(\frac{r^2}{r_1 r_2}\right)\right] \end{aligned} \tag{6.7}$$

where I_{app} is the approximate sound intensity obtained by the two-microphone intensity method, I_{ex} is the analytical exact sound intensity, r,

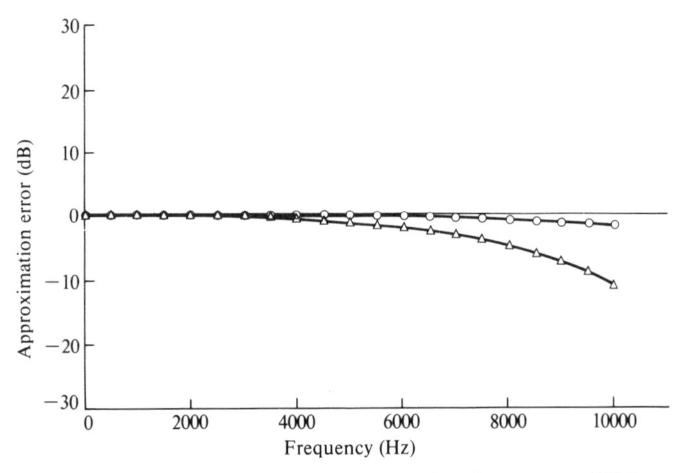

Fig. 6.5. Examples for monopole-source approximation errors (With permission from Journal of Sound and Vibration. Copyright: Academic Press (London) Inc.).[6.9] $r = 24$ mm; (○) $r_2 - r_1 = 8$ mm; (△) $r_2 - r_1 = 16$ mm.

Δr, r_1, and r_2 are defined in Fig. 6.4, and k is the wave number, i.e. $k = 2\pi f/c$, where c is the speed of sound. As an illustration, the above expression is plotted in Fig. 6.5 over a frequency range from 0 to 10 kHz, with a measurement distance r of 24 mm and two microphone separation Δr values of 8 and 16 mm. Figure 6.5 shows that the errors are within ± 1 dB for frequencies up to 5.7 kHz with $\Delta r = 16$ mm or up to 11 kHz with $\Delta r = 8$ mm.

For a lateral quadrupole source, the finite-difference approximation error can be expressed as[6.9]

$$L_e = 10 \log_{10}\left[\left(\frac{\sqrt{(9+3k^2r_1^2+k^4r_1^4)}\,\sqrt{(9+3k^2r_2^2+k^4r_2^4)}}{k^5\,\Delta r r_1^2 r_2^2}\sin(k\,\Delta r+\alpha_1-\alpha_2)\right)\times\left(\frac{r^2}{r_1 r_2}\right)\right] \quad (6.8)$$

where

$$\alpha_1 = \arctan[(k^2r_1^2-3)/3kr_1]$$

and

$$\alpha_2 = \arctan[(k^2r_2^2-3)/3kr_2].$$

The above expression is plotted in Fig. 6.6, using the same parameters as for Fig. 6.5. Large errors are seen in Fig. 6.6 for frequencies below 500 Hz.

As the near field due to an actual noise source is often much more complex than that of the monopole, it is not appropriate to approximate

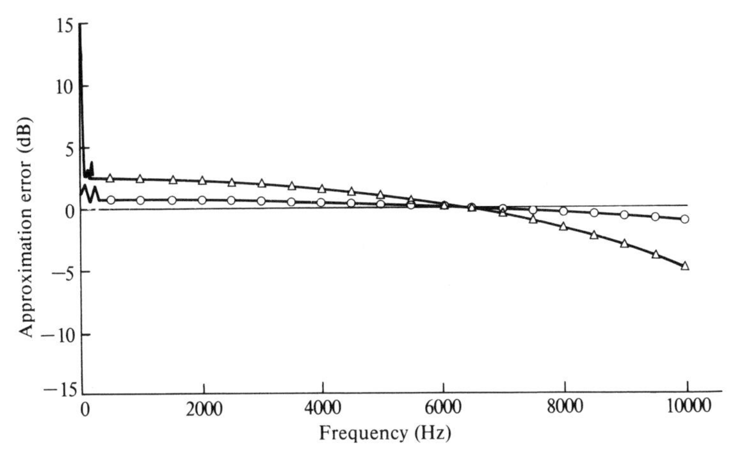

Fig. 6.6. Examples for quadrupole-source approximation errors (With permission from Journal of Sound and Vibration. Copyright: Academic Press (London) Inc.).[6.9] $r = 24$ mm; (○) $r_2 - r_1 = 8$ mm; (△) $r_2 - r_1 = 16$ mm.

actual sources as monopoles in near-field intensity measurements. It has been suggested[6.9] that, for an actual-source near-field intensity measurement, the best procedure would be to select parameter values, i.e. $k\,\Delta r$ and $\Delta r/r$ values, based on the quadrupole analysis as a 'worst case'. The following parameter limits would be useful guidelines for an accuracy of ±1.5 dB.

$$0.1 \leqslant k\,\Delta r \leqslant 1.3 \tag{6.9}$$

and

$$0 \leqslant \frac{\Delta r}{r} \leqslant 0.5. \tag{6.10}$$

For a given Δr value, the lower frequency limit can thus be determined by $k\,\Delta r = 0.1$ and the upper frequency limit by $k\,\Delta r = 1.3$. Table 6.1 shows the useful frequency ranges based on eqn (6.9).

TABLE 6.1. Useful frequency range for two-microphone sound-intensity measurement based on a quadrupole source with an accuracy of ±1.5 dB

Δr (mm)	Lower frequency limit (Hz)	Upper frequency limit (kHz)
6	900	11.7
12	450	5.86
24	225	2.93
48	112	1.46

Another error source in the two-microphone intensity method is the phase mismatch between the two microphone channels and this is particularly important at low frequencies. The error in sound intensity due to the phase mismatch can be expressed as[6.10]

$$L_e = 10 \log_{10}\left(\frac{I_{app}}{I_{ex}}\right)$$

$$\approx 10 \log_{10}\left(1+\frac{\theta_e}{\theta_p}\right) \quad \text{for} \quad \theta_e \ll 1 \quad \text{and} \quad \theta_p \ll 1 \qquad (6.11)$$

where θ_e is the instrument phase mismatch in radians between the two microphone channels and θ_p is the physical phase shift in radians between the pressure signals at the two measuring points for a given frequency. According to eqn (6.11), at 100 Hz with $\Delta r = 10$ mm, a small phase mismatch of one degree between the two channels gives an L_e value of approximately 3 dB. Thus, it is essential to keep the instrument phase mismatch to a minimum. The use of two phase-matched channels or a phase calibration between the two entire measurement channels would be helpful. However, these are not practicable for routine field measurements.

Chung[6.10] has therefore developed a circuit-switching technique to eliminate the instrument phase-mismatch error. The switching technique requires an interchange of the entire first and the second microphone instrumentation channels, including the microphone cartridges, amplifiers, and filters, and the determination of the cross-spectrum of $p_1(t)$ and $p_2(t)$ under original and switched conditions. The sound intensity can be calculated from

$$I_r(\omega) = \mathrm{Im}[(G_{12}G_{12}^S)^{1/2}]/(2\rho\omega\,\Delta r\,|H_1|\,|H_2|) \qquad (6.12)$$

where $\mathrm{Im}[\cdots]$ means the imaginary part of $[\cdots]$, G_{12} is the cross-spectrum between $p_1(t)$ and $p_2(t)$ under original test conditions, G_{12}^S is the cross-spectrum under switched test conditions, and $|H_1|$ and $|H_2|$ are the gain factors of the two microphone channels. Readers should refer to Section 7.5 for the definition of cross-spectrum.

Although the two-microphone sound-intensity method has a rather limited frequency range, it is particularly useful in determining the sound power of a machine under actual operating conditions. Laville *et al.*[6.11] have used the method to measure the sound power of a steam valve in a power station. This is based on the fact that the sound intensity in a given direction is a signed quantity. For example, the sound intensity on surface 1 (see Fig. 6.7) in the $\mathbf{n}_1$ direction due to the source B on the left of the closed surface is negative since the sound power flows in the direction

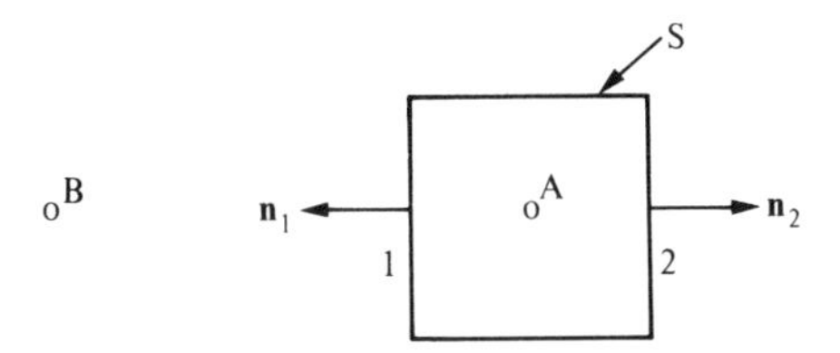

Fig. 6.7. Noise sources inside and outside a closed surface. S—Closed surface; A—source inside S; B—source outside S; $\mathbf{n}_1$—outward direction normal to surface 1; $\mathbf{n}_2$—outward direction normal to surface 2.

opposite to $\mathbf{n}_1$. The total power emitting from the closed surface due to the source B is therefore zero, based on the integration of $I_\mathbf{n}\,\mathrm{d}A$ over the closed surface. This is true on condition that there is no flow, the absorption within the closed surface is negligible, and the major noise components of the source A enclosed by the surface are different from those of the source B situated outside the surface.

6.2.3. *Surface intensity method*

On the surface of a machine one can measure the sound pressure with a microphone situated very close to the surface and measure the velocity with a vibration transducer fixed to the surface. The sound intensity on the surface can be determined by the system shown in Fig. 6.3, according to eqn (6.1).

Robust and inexpensive accelerometers can be used to measure the surface vibration in the normal direction of the machine surface. Hodgson and Erianne[6.7] used a 2-gm accelerometer fixed to the machine and a 0.66-cm microphone located very close (1 mm) to both the machine surface and the accelerometer. The signals from the accelerometer and microphone were first tape-recorded and then digitized for computer processing to calculate the surface acoustic intensity for a particular frequency band. The total sound-power emission for a given frequency band can be found from multiple-point measurements of surface acoustic intensity, based on eqn (6.3).

A number of researchers[6.12]–[6.13] have studied the accuracy of the surface-intensity method and found that it was essential to take into account the phase shifts between the pressure and surface vibration signals. The phase shifts are introduced not only by the instrument phase mismatch between the vibration and pressure channels, but also by the finite distance between the microphone and the surface on which the vibration transducer is placed. If the perpendicular distance between the diaphragm of the microphone and the vibrating surface is ΔX, the phase

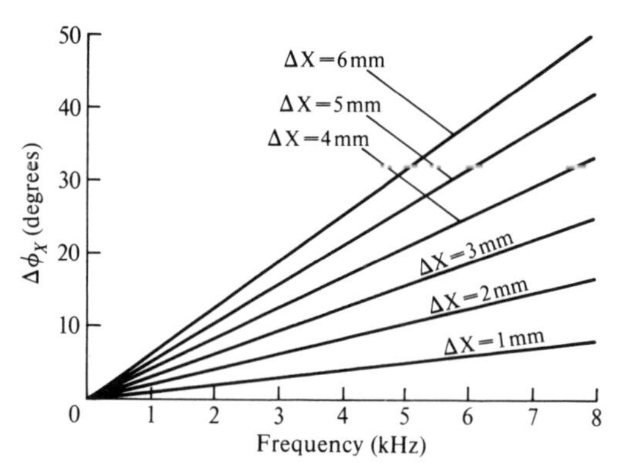

Fig. 6.8. Phase shift $\Delta\phi_X$ due to the finite distance ΔX between the microphone and the vibrating surface.[6.13]

shift in degrees can be expressed by

$$\Delta\phi_X = 360f\,\Delta X/c. \tag{6.13}$$

Equation (6.13) is plotted in Fig. 6.8 for a number of ΔX values over the frequency range from 0 to 8 kHz. Figure 6.8 shows clearly that a small ΔX value of 1 mm can introduce a significant phase shift at high frequencies.

Kaemmer and Crocker[6.13] studied the error in surface intensity in terms of the phase shifts and the cross-spectral density between surface velocity and pressure. It can be expressed as

$$L_e = 10\log_{10}\left(\frac{I}{I_t}\right) = -10\log_{10}[\cos\phi + (Q_{vp}/C_{vp})\sin\phi] \tag{6.14}$$

where I_t is the true surface intensity when the phase shifts have been taken into account, I is the approximate surface intensity neglecting the effect of the phase shifts, ϕ is the phase shift angle, Q_{vp} is the quad-spectrum (the imaginary part), and C_{vp} is the co-spectrum (the real part) of the one-sided cross-spectral density between surface velocity and pressure (see Section 7.5 for the definition of co-spectrum and quad-spectrum).

As surface-intensity measurements can give the local surface intensity values at various parts of a machine, the measurements are particularly useful in determining the predominant surface areas for noise emission.

References

[6.1] van Zyl, B. G. and Anderson, D. (1975). Evaluation of the intensity method of sound power determination. *J. acoust. Soc. Am.* **57,** 682–6.

[6.2] Baker, S. (1975). An acoustic intensity meter. *J. acoust. Soc. Am.* **27,** 269–73.
[6.3] Schultz, T. (1956). Acoustic wattmeter. *J. acoust. Soc. Am.* **28,** 693–9.
[6.4] Olson, H. F. (1974). Field-type acoustic wattmeter. *J. Audio Eng.* **22,** 321–7.
[6.5] Olsen, H. F. (1972). *Modern sound reproduction.* Van Nostrand-Reinhold, New York.
[6.6] Olsen, H. F. (1957). *Acoustical engineering.* Van Nostrand-Reinhold, New York.
[6.7] Hodgson, T. H. and Erianne, R. D. (1976). Noise source location for large machines from measurements of surface acoustic intensity. *Proceedings of Inter-Noise 76*, pp. 13–18. Institute of Noise Control Engineering, New York.
[6.8] Fahy, F. J. (1977). Measurement of acoustic intensity using the cross-spectral density of two microphone signals. *J. acoust. Soc. Am.* **62**(4), 1057–9.
[6.9] Thompson, J. K. and Tree, D. R. (1981). Finite difference approximation errors in acoustic intensity measurements. *J. Sound Vibration* **75**(2), 229–38.
[6.10] Chung, J. Y. (1978). Cross-spectral method of measuring acoustic intensity without error caused by instrument phase mismatch. *J. acoust. Soc. Am.* **64**(6), 1613–16.
[6.11] Laville, F., Salvan, G., and Pascal, J. C. (1980). Sound power determinations using intensity measurements under field conditions. *Proceedings of Inter-Noise 80*, pp. 1083–6. Noise Control Foundation, New York.
[6.12] Czarnecki, S., Engel, Z., and Panuszka, R. (1976). Correlation method of measurements of sound power in the near field conditions. *Arch. Acoustics* **1**(3), 201–13.
[6.13] Kaemmer, N. and Crocker, M. J. (1979). Sound power determination from surface intensity measurements on a vibrating cylinder. *Proceedings Noise-Con 79*, pp. 153–60. Noise Control Foundation, New York.

Further reading for Chapter 6

[1] Crocker, M. J. (1981). Determination of noise sources and sound power using the surface intensity and acoustic intensity approach. *Proceedings of Inter-Noise 81*, pp. 895–8. Nederlands Akoestisch Genootschap, Delft.
[2] Elliot, S. J. (1981). Spatial sampling of acoustic intensity fields. *Proceedings of Inter-Noise 81*, pp. 899–902. Nederlands Akoestisch Genootschap, Delft.
[3] Laville, F. (1981). A comparative study of intensity versus pressure measurements. *Proceedings of Inter-Noise 81*, pp. 903–6. Nederlands Akoestisch Genootschap, Delft.
[4] Rasmussen, G. and Brock, M. (1981). Noise measurement, analysis and instrumentation, acoustic intensity measuring probe. *Proceedings of Inter-Noise 81*, pp. 907–12. Nederlands Akoestisch Genootschap, Delft.
[5] Bristo, J. D. (1979). Machinery noise source analysis using surface intensity measurements. *Proceedings of Noise-Con 79*, pp. 137–42. Noise Control Foundation, New York.
[6] Hickling, R. (1981). Sound power measurement using the cross-spatial technique. *Proceedings of Noise-Con 81*, pp. 25–30. Noise Control Foundation, New York.

[7] Lambrich, H. P. and Stahel, W. A. (1977). A sound intensity meter and its applications in car acoustics. *Proceedings of Inter-Noise 77*, B142–7. International Institute of Noise Control Engineering, Zürich.
[8] Hübner, G. (1982). Analysis of errors and field of application for sound intensity measurements for the determination of sound power of sound sources under 'in situ' measurement conditions. *Proceedings of Inter-Noise 82*, pp. 691–4. Noise Control Foundation, New York.
[9] Kiteck, P. and Tichy, J. (1982). Intensity probe obstacle effects and errors of the transfer function technique to calibrate acoustic intensity measurements. *Proceedings of Inter-Noise 82*, pp. 695–8. Noise Control Foundation, New York.
[10] Hübner, G. (1983). Determination of sound power of sound sources under in-situ conditions using intensity method-field of application, suppression of parasitic noise, reflection effect. *Proceedings of Inter-Noise 83*, pp. 1043–6. Institute of Acoustics, Edinburgh.

7. Noise measuring equipment

7.1. Microphones

All noise measurements are based on the use of a microphone to convert acoustical energy to electrical energy for further analysis. At present most microphones commercially available are condenser, electret, or piezoelectric microphones. The condenser microphone is mainly used for precision noise measurement while the piezoelectric microphone, cheaper and less stable, is used for general-purpose noise measurement. The electret microphone has been widely used for non-professional purposes, e.g. in cassette-recorders, radio sets, and public address systems, and has recently increased its share in both precision and general-purpose noise measurements.

7.1.1. *Condenser microphones*

A schematic diagram of a condenser microphone is given in Fig. 7.1. It consists essentially of a thin metallic diaphragm and a rigid backplate. The diaphragm and backplate are electrically insulated from each other and connected to a stabilized direct-voltage polarization source, forming a parallel-plate capacitor. When the microphone is exposed to a sound-pressure wave, the thin diaphragm moves to and fro relative to the rigid backplate. This movement causes an alternating change in the capacitance between the diaphragm and the backplate and hence produces a corresponding voltage signal across the output terminals.

The sensitivity of a condenser microphone depends mainly on the polarization voltage, ambient pressures, diaphragm area, and diaphragm tension. An example of the sensitivity variation with ambient pressure for a condenser microphone is shown in Fig. 7.2.[7.1] Typical open-circuit sensitivity is $50\,\mathrm{mV\,N^{-1}\,m^2}$ for a one-inch condenser microphone, $12.5\,\mathrm{mV\,N^{-1}\,m^2}$ for a half-inch condenser microphone, 1.5 to $4\,\mathrm{mV\,N^{-1}\,m^2}$ for a quarter-inch condenser microphone, and $1\,\mathrm{mV\,N^{-1}\,m^2}$ for an eighth-inch condenser microphone. The approximate dynamic range is from 20 to 140 dB for a one-inch condenser microphone, from 25 to 160 dB for a half-inch one, from 42 to 170 dB for a quarter-inch one, and from 50 to 178 dB for an eighth-inch one.

7.1.2. *Electret microphones*

During the last decade electret microphones suitable for both precision and general-purpose noise mesurement have become commercially available. An electret is an electrically polarized element which can maintain

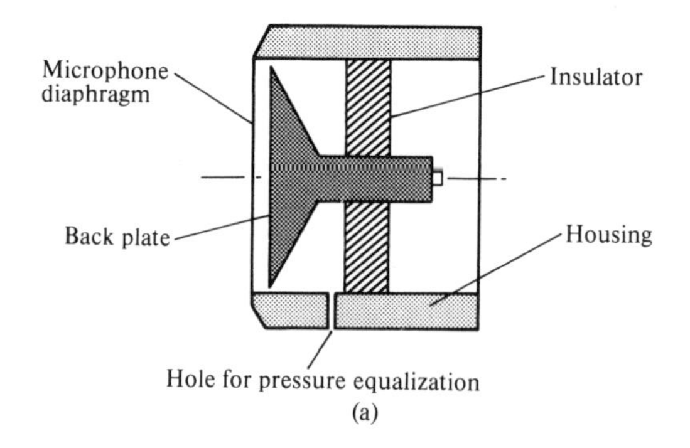

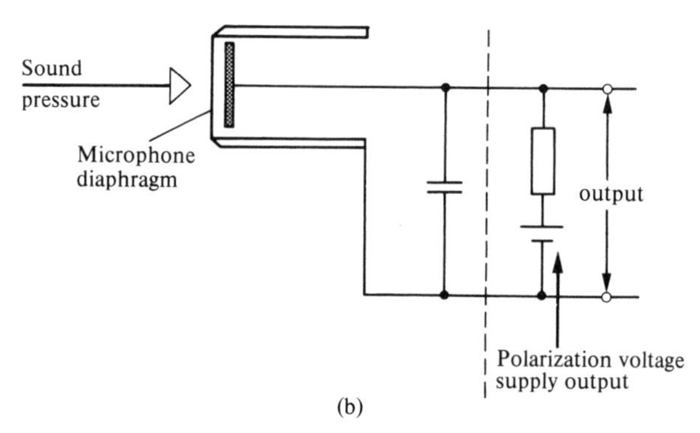

Fig. 7.1. Schematic diagram of a condenser microphone. (a) Microphone diaphragm and back plate; (b) microphone and polarization voltage supply.

its charge and polarization for a very long period. Figure 7.3(a) shows the typical design of a mass-produced electret microphone.[7.2] The microphone diaphragm is a metallized plastic foil which has been specially treated and charged in such a way that there is an electrical potential between the inside surface of the plastic foil and the metallized outside surface. Figure 7.3(b) shows the frozen charge on the inside surface of the foil and the image charge on the outside metal surface and on the backplate. The frozen charge and the charge on the backplate set up an electric field between the foil and the backplate similar to the electric field set up by the external direct polarization voltage of a condenser microphone. Therefore electret microphones are also called prepolarized condenser microphones.

The acoustic characteristics of the electret microphone are approximately the same as those of the condenser microphone. However, in comparison with the condenser microphone, the electret microphone has

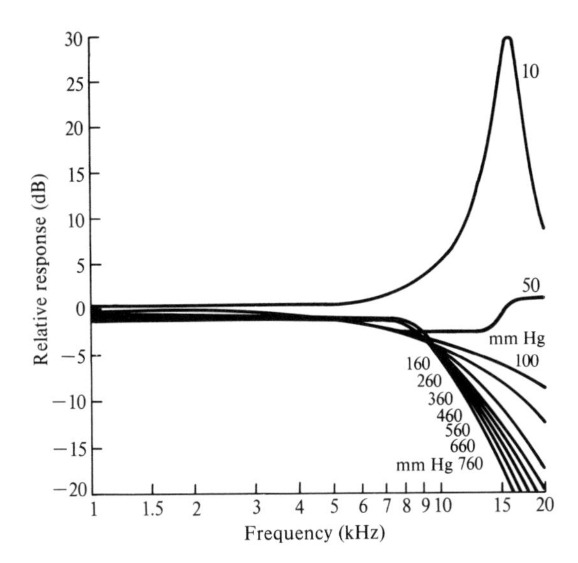

Fig. 7.2. Effect of ambient pressure on frequency response of a half-inch condenser microphone (Reproduced by permission of Brüel & Kjaer).[7.1] (760 mm Hg = 1 atm.)

the following advantages:

(1) It does not need an external direct polarization voltage supply and hence the complete noise measuring equipment can be made lighter and smaller;
(2) It is more robust mechanically;
(3) It is more reliable in a humid environment.

7.1.3. *Piezoelectric microphones*

A schematic diagram of a piezoelectric microphone is given in Fig. 7.4. When the sound pressure deflects the diaphragm, the movement of the diaphragm induces a deformation in the piezoelectric plate which in turn generates a voltage signal across the output terminals. The most common piezoelectric materials used in microphones are lead zirconate titanate, barium titanate, and Rochelle salt. As piezoelectric materials are sensitive to temperature and humidity changes, they have rather limited operating temperature and humidity ranges. Piezoelectric microphones using lead zirconate titanate can, however, be used with confidence over a temperature range of −10 °C to +50 °C and in an environment having a relative humidity up to 90 per cent. The operating frequency range is usually confined to the range from about 32 Hz to 8 kHz. Owing to the limited operating frequency range and long-term stability problems, piezoelectric

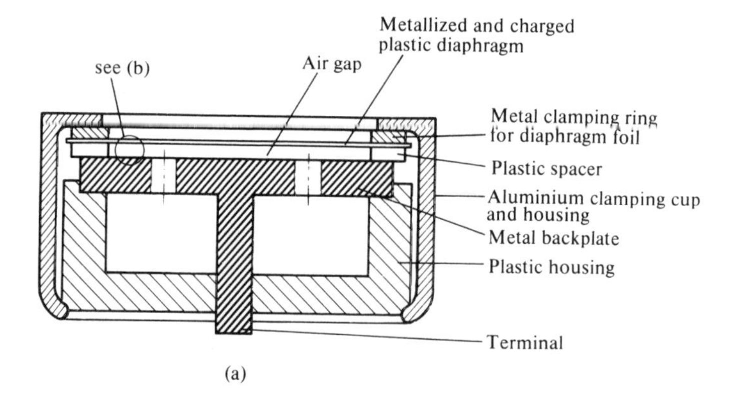

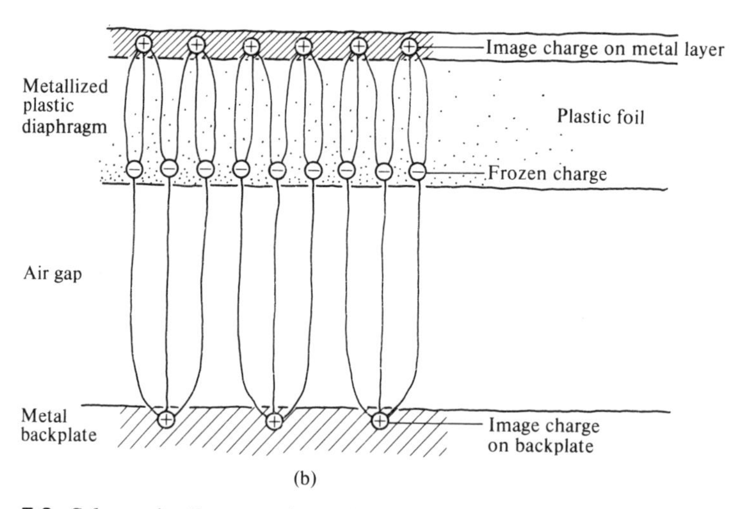

Fig. 7.3. Schematic diagram of an electret microphone (Reproduced by permission of Brüel & Kjaer).[7.2] (a) Basic parts; (b) charge distribution.

microphones are used only in general-purpose industrial noise-measuring equipment.

7.1.4. *Microphone frequency response*

When a sound wave impinges on the diaphragm of a microphone, it will be partly reflected from it, causing an increase in the sound pressure at the diaphragm. This pressure increase due to reflection becomes signific-

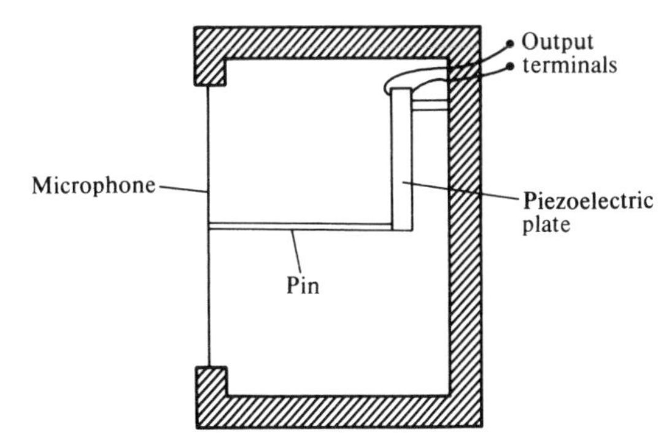

Fig. 7.4. Schematic diagram of a piezoelectric microphone.

ant for sound waves having frequencies above 1 kHz. The pressure increase is usually determined experimentally by the microphone manufacturer for different angles of incidence (see Fig. 7.5). Figure 7.6 shows the pressure increase expressed in dB for a one-inch condenser microphone. The values of the pressure increase in dB are called free-field corrections. In addition to the free-field corrections, the microphone manufacturer should also determine the pressure response of a microphone, using the reciprocity calibration. The reciprocity calibration is an absolute calibration method and its procedures and principles are described in IEC 327,[7.3] IEC 402,[7.4] and ANSI Sl.10-1966.[7.5] The accuracy obtained in the pressure response is believed to be about 0.05 dB.

The free-field frequency response is obtained by adding the free-field corrections to the pressure response. It is easy to verify that the free-field frequency-response curve for a one-inch condenser microphone at 0° incidence given in Fig. 7.7 for the B & K Type 4145/61 microphone is the sum of the pressure-response curve (curve P) and the free-field correction curve for 0° incidence of Fig. 7.6.

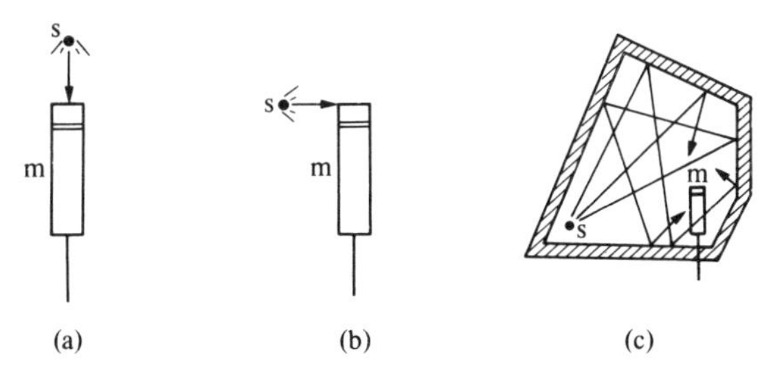

Fig. 7.5. Angles of incidence. (a) 0° incidence; (b) 90° incidence; (c) random incidence. s = source; m = microphone.

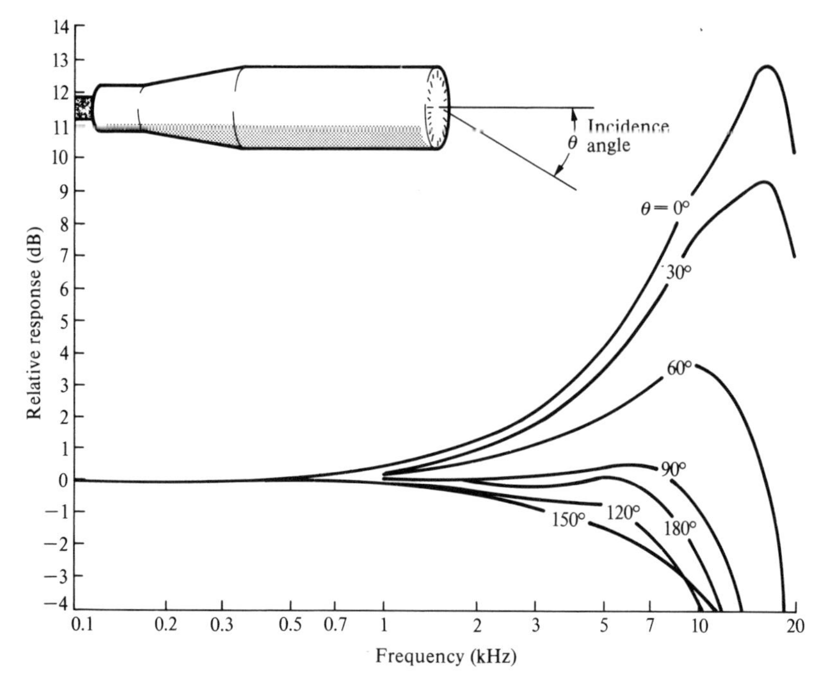

Fig. 7.6. Free-field corrections of one-inch condenser microphone at various angles of incidence (Reproduced by permission of Brüel & Kjaer).[7.1]

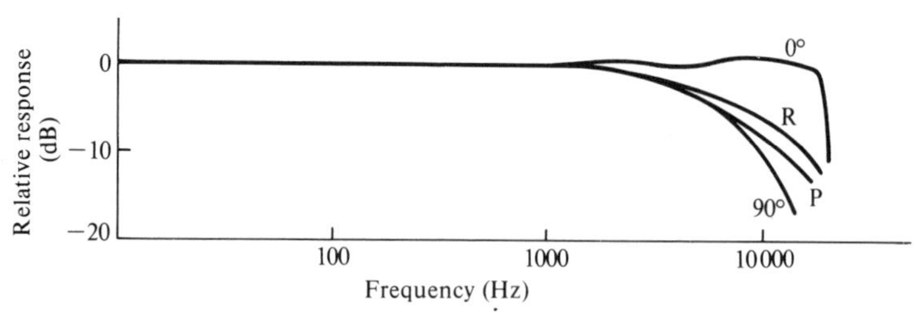

Fig. 7.7. Free-field frequency response of one-inch condenser microphone (Reproduced by permission of Brüel & Kjaer).[7.1] 0° = at 0° incidence; 90° = at 90° incidence; R = random incidence; P = pressure response curve.

7.2. Sound level meters

A sound level meter is a measuring device which converts the sound pressure variation into a meter reading corresponding to the combined weighted sound pressure level in the entire audible frequency range with A-weighting adjustments. This reading, described in Section 1.7, is called the sound level in A-weighting. Many commercial sound level meters are

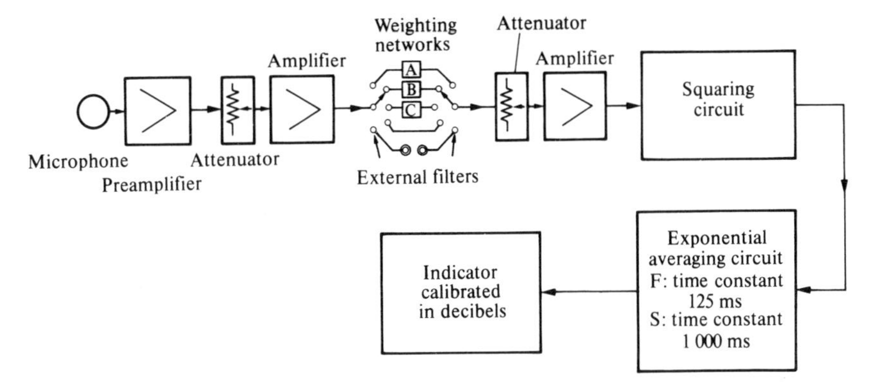

Fig. 7.8. Schematic diagram of a sound-level meter.

also equipped with a linear frequency response and some have B- and C-weightings (see Chapter 1).

7.2.1. *Analogue sound level meters*

A schematic diagram of an analogue sound level meter is shown in Fig. 7.8. The meter consists of the following main sections: microphone and impedance matching device; input attenuator and amplifier; weighting networks; output attenuator and amplifier; and detector-indicator. A simplified diagram of A-, B-, and C-weighting networks is given in Fig. 7.9. These weighting networks should provide the relative frequency response shown in Fig. 1.4.

The tolerance limits, i.e. the accuracy, for a sound level meter are given in IEC 651[7.15] and its equivalent BS 5969.[7.16] Prior to the publication of IEC 651 in 1979, the accuracy for a precision sound level meter had been specified in IEC 179[7.6], BS 4197,[7.7] and ANSI S1.4–1971 (Type 1).[7.8] The accuracy of an industrial grade sound level meter had been specified by IEC 123[7.9] and BS 3489.[7.10]

IEC 651 specifies the accuracy of sound level meters with four degrees of precision, designated types 0, 1, 2, and 3. The type-0 sound level meter is intended as a laboratory reference standard. Type 1 is for laboratory use and for field use where the acoustical environment can be

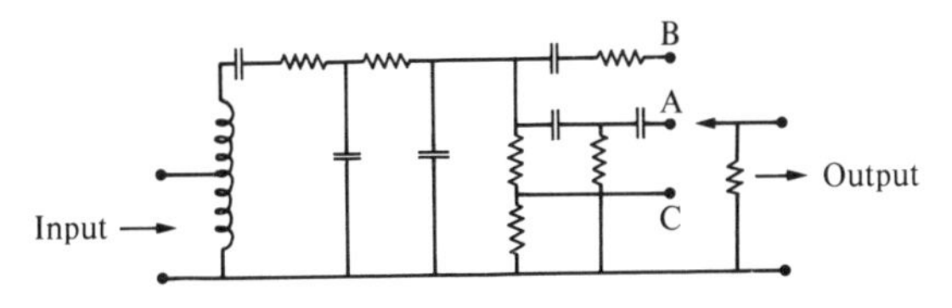

Fig. 7.9. Simplified A-, B-, and C-weighting networks.

TABLE 7.1. Response to tone bursts of F (fast) and S (slow) detector–indicators[7.15]*

Detector–indicator characteristic	Duration of test tone burst (ms)	Maximum response to test-tone burst referred to response to continuous signal (dB)	Tolerances on maximum response for each instrument type (dB)			
			0	1	2	3
	Continuous	0	—	—	—	—
F	200	−1.0	±0.5	+1	+1 −2	+1 −3
	50	−4.8	±2	—	—	—
	20	−8.3	±2	—	—	—
	5	−14.1	±2	—	—	—
S	2000	−0.6	±0.5	—	—	—
	500	−4.1	±0.5	±1	±2	±2
	200	−7.4	±2	—	—	—
	50	−13.1	±2	—	—	—

* (Reproduced by permission of the International Electrotechnical Commission, which retains the copyright.)

closely specified and/or controlled. The type-2 sound level meter is suitable for general purpose field applications and type 3 is primarily for field noise survey applications.

The detector–indicator of a sound level meter has two different characteristics: F (fast) and S (slow). According to IEC 651, the characteristics of the detector–indicator shall be such that when the applied signal is suddenly turned off, the indicator will decay by 10 dB in a time of 0.5 s or less for 'F' and in a time of 3.0 s or less for 'S'. The response to tone bursts of both 'F' and 'S' characteristics is shown in Table 7.1.

'Slow' response smooths the meter readings and is suitable for measuring steady machinery noise and factory noise. 'Fast' response follows short-duration noise variations more quickly and is used for measuring unsteady machinery noise.

7.2.2. *Impulse sound-level meters*

Impulse sound-level meters are equipped with I (i.e. Impulse) detector–indicators for measuring impulsive noise, e.g. punch press noise. Figure 7.10 shows the I detector–indicator characteristics in accordance with IEC 651[7.15] and its equivalent BS 5969.[7.16] The impulse sound-level meter has a short rise-time constant and a very long decay-time constant. The averaging-circuit time constant for the impulse mode is 35 ms,

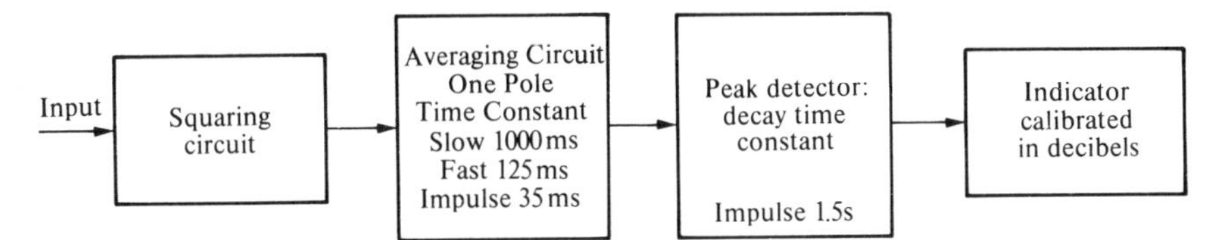

Fig. 7.10. Impulse detector–indicator characteristics (Reproduced by permission of the International Electrotechnical Commission, which retains the copyright).[7.15]

compared with 125 ms for the Fast mode and 1000 ms for the Slow mode. All commercial impulse sound-level meters also have F (fast) and S (slow) detector–indicators and the maximum responses of the meters in the three modes to pulses of different durations are shown in Fig. 7.11. The asymptotic differences in the three modes are[7.17]

$$L_{imp} - L_{fast} = 5.5 \text{ dB},$$
$$L_{fast} - L_{slow} = 9 \text{ dB},$$
$$L_{imp} - L_{slow} = 14.5 \text{ dB}.$$

Figure 7.12 shows the time history of the indications in the three modes when a single pulse of 50-ms duration is applied to the input of the meter. Owing to the large decay-time constant of 1500 ms in its peak detector the impulse mode results in the slowest decay. According to IEC 651,[7.15] indications in the slow, fast, and impulse detector–indicator modes shall not differ by more than 0.1 dB for types 0, 1, and 2 and 0.2 dB for type 3 meters for steady-state sinusoidal signals in the frequency range 315 Hz–8 kHz. However, when tested with a single burst or a continuous sequence of bursts with low repetition rate, the Impulse

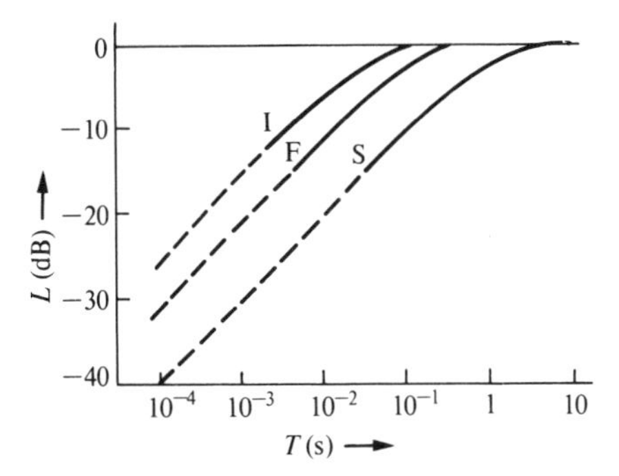

Fig. 7.11. Response of the impulse (I), fast (F), and slow (S) modes to pulses of duration T (Reproduced by permission of the Institute of Noise Control Engineering, which retains the copyright).[7.17]

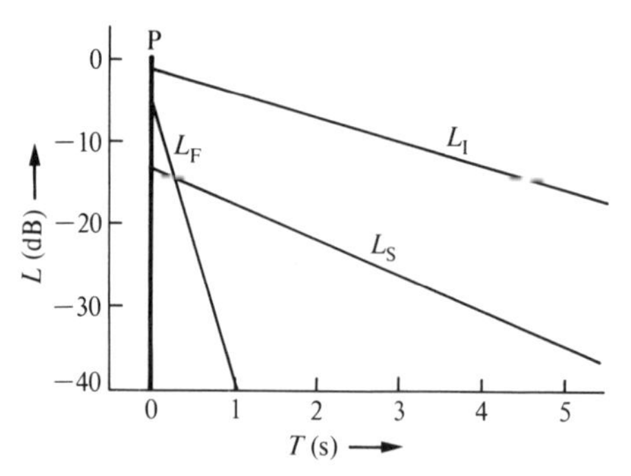

Fig. 7.12. Time history of levels in impulse, fast, and slow modes in response to a pulse P of 50-ms duration (Reproduced by permission of the Institute of Noise Control Engineering, which retains the copyright).[7.17] L_I = level in impulse mode; L_F = level in fast mode; L_S = level in slow mode.

mode will generally give an indication that is higher than either Fast or Slow. Optionally, the sound-level meter may be equipped for measuring peak values. In the Peak mode, for a type 0 meter, the onset time of the detector shall be such that a single pulse of 50-μs duration produces a deflection no more than 2 dB below that produced by a pulse having a duration of 10 ms and equal peak amplitude. This requirement should be met for pulses of both polarities. For type 1, 2, and 3 meters, the onset time should be such that a single 100-μs pulse of either polarity produces a deflection no more than 2 dB below the deflection produced by a pulse having a duration of 10 ms and equal peak amplitude. This allows the peak values of a sound whose duration may be as short as 50 to 100 μs to be recorded reasonably accurately. As an example, Table 7.2 shows the readings obtained by Brüel[7.18] on impulse sound-level meters for various noise sources in impulse, peak, and fast modes. It is of interest to note that, for a 40-ton punch press, the sound level near the operator's head is 121 dB(A) in peak mode, compared with 98 dB(A) in impulse mode and 93 dB(A) in fast mode.

7.2.3. *Digital sound-level meters and integrating sound-level meters*

In recent years many digital sound-level meters have become available commercially. The simplest digital sound-level meter consists of a conventional analogue sound-level meter and a digital display. The more sophisticated digital sound-level meter can calculate the equivalent continuous sound level L_{eq} for a predetermined period and is called an integrating sound-level meter. Figure 7.13 shows a schematic diagram of an integrating sound-level meter. The digital signal from the analogue-to-digital converter is fed to a gate which is controlled by the measurement-period selector. The measurement period can be selected within a range

TABLE 7.2. Readings on impulse sound-level meters for various noise sources in fast, impulse, and peak modes[7.18]

Sound source	Fast (dB(A))	Imp. (dB(A))	Imp. hold (dB(A)) 5×	Peak hold (dB(A)) 5×	Δ*
Sinusoidal pure tone 1000 Hz	94	94	94	97	3
Beat music from a gramophone	90	91	93	97	4
Modern music from a gramophone	102	103	103	105	2
Electric guitar from a gramophone	85	86	86	91	5
Motorway traffic 15 m distance	80	80	81	89	8
Motorway traffic 50 m distance	68	68	68	76	8
Train 70 km/h rail noise 10 m distance	95	96	98	106	8
Train 70 km/h rail noise 18 m distance	85	87	87	94	7
Noise in aircraft type PA 23, cruising speed	90	91	91	100	9
Noise in aircraft type *Falco* F 8, crusing speed	97	98	98	109	11
Noise in aircraft type *KY* 3, cruising speed	102	102	103	112	9
Noise in car type *Fiat* 500, 60 km/h	78	79	79	93	14
Noise in car type *Volvo* 142, 80 km/h	75	75	76	86	10
Lawn mower 10 HP 1 m distance	97	99	99	116	17
Typewriter IBM (Head position)	80	84	83	102	19
Electric shaver 2.5 m distance	92	92	92	107	15
75 HP diesel motor in electricity generating plant	100	101	101	113	12
Pneumatic nailing machine 3 m distance	112	114	113	128	15
Pneumatic nailing machine near operator's head	116	120	120	148	28
Industrial ventilator 5 HP 1 m	82	83	83	93	10
Air compressor room	92	92	92	104	12
Large machine shop	81	82	82	98	16
Turner shop	79	80	81	100	19
Automatic turner shop	79	80	80	99	19
40 tons punch press, near operator's head	93	98	97	121	24
Small automatic punch press	100	103	103	118	15
Numerically driven high-speed drill	100	102	103	112	9
Small high-speed drill	98	101	101	109	8
Ventilator with filter	82	83	83	94	11
Machine driven saw, near operator's head	102	102	104	113	9
Vacuum cleaner type *Hoover*, 1.2 m distance	81	81	81	93	12
Bottles striking each other	85	88	90	105	15
Bottling machine in brewery	98	99	101	122	21
Toy pistol (cap)	105	108	108	140	32
Pistol 9 mm, 5 m distance from side	113	114	116	146	30
Shotgun 5 m distance from side	108	110	111	143	32
Saloon rifle 1 m distance from side	107	110	110	139	29

* Δ = Peak hold − Imp. hold.

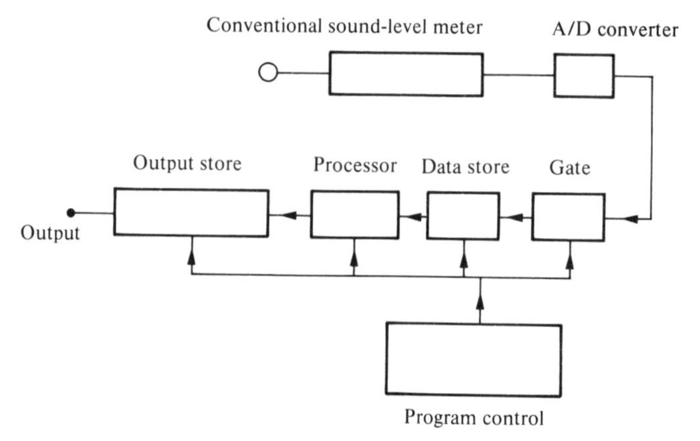

Fig. 7.13. Schematic diagram of integrating sound level meter.

from a few minutes to a few hours or more. During the selected period the gate will open, typically 1024 times, each time admitting an instantaneous sample of the sound level to the data memory. At the end of the selected period the 1024 samples are made available to the processor which will then calculate the equivalent continuous sound level L_{eq} based on eqn (2.6). The resulting L_{eq} value will be put into the answer memory and the result can be shown on a digital display or can be fed to a printer or recorder. With the aid of the program control, the processor can also calculate the L_{10} and L_{90} values mentioned in Section 2.5.

7.3. Frequency analysers

A frequency analyser measures the frequency spectrum of a noise. It has the same main parts as a sound-level meter except that it is equipped with a filter set in addition to the weighting networks. Therefore, a sound-level meter can become a frequency analyser simply by adding a filter set. The filter set may consist of octave-band filters, one-third octave-band filters, narrow-band constant-percentage bandwidth filters, or narrow-band constant-bandwidth filters. A frequency analyser equipped with only octave-band filters is called an octave analyser.

7.3.1. *Analogue narrow-band frequency analysers*

An analogue narrow-band frequency analyser is equipped with a set of narrow-band filters having typically selectable constant-percentage bandwidths of 1, 3, 10, and 23 per cent or selectable constant bandwidths of 3.16, 10, 31.6, 100, 316, and 1000 Hz.

The filter is a continuously variable bandpass filter and can be tuned to any frequency within the audible-frequency range manually or by an external mechanical drive. The output from the analyser can be read from the indicating meter or fed to a level recorder to plot a frequency spectrum. The constant-percentage bandwidth analyser provides a better resolution at low frequencies while the constant-bandwidth analyser can give a better resolution at high frequencies (see Fig. 1.2).

7.3.2. *Digital real-time frequency analysers (single-channel)*

All digital and microprocessor-based frequency analysers are now produced by a number of noise-measuring equipment manufacturers. These digital frequency analysers can produce and display a complete frequency spectrum in a very short period, e.g. less than 0.5 second; it may take a few minutes or more to obtain a complete frequency spectrum using an analogue frequency analyser together with a level recorder. A digital frequency analyser may be used for 'real-time frequency analysis', which means 'analysis of all signals in all frequency bands at all times, displaying the results on a continuously updated screen'. There are two different digital techniques for frequency analysis—by digital filtering and by the use of fast Fourier transforms.

A digital filter is a digital processor which consists of a number of multipliers and adders and can perform a filtering action on digital signals equivalent to that of an analogue filter operating on the equivalent continuous analogue signals.

Figure 7.14 shows the block diagram of a two-pole digital filter. The principles of digital filtering and the design of digital filters are beyond the scope of this book and are discussed elsewhere.[7.11],[7.12] However, the basic features of a digital filter will now be described.

In Fig. 7.14, the filter coefficients of A_0, A_1, A_2, B_1, and B_2 determine completely the shape of the amplitude responses of the filter. However, the centre frequency and passband bandwidth are directly proportional to the sampling frequency. Figure 7.15 shows an example of the amplitude response curve of a bandpass digital filter and it should be noted that the horizontal axis is in normalized frequency, defined as the ratio of frequency to $F_s/2$, where F_s is the sampling frequency. Thus, doubling the sampling frequency and using the same filter coefficients would result in a new filter, having the same shape of amplitude response but operating with a new centre frequency and a new passband bandwidth which are twice the original values. A digital filter having a set of fixed filter coefficients is a constant-percentage bandwidth filter and can be made to represent an infinite number of different filters simply by varying the sampling frequency. Also, the filter amplitude response can be changed easily by programming in a new set of filter coefficients.

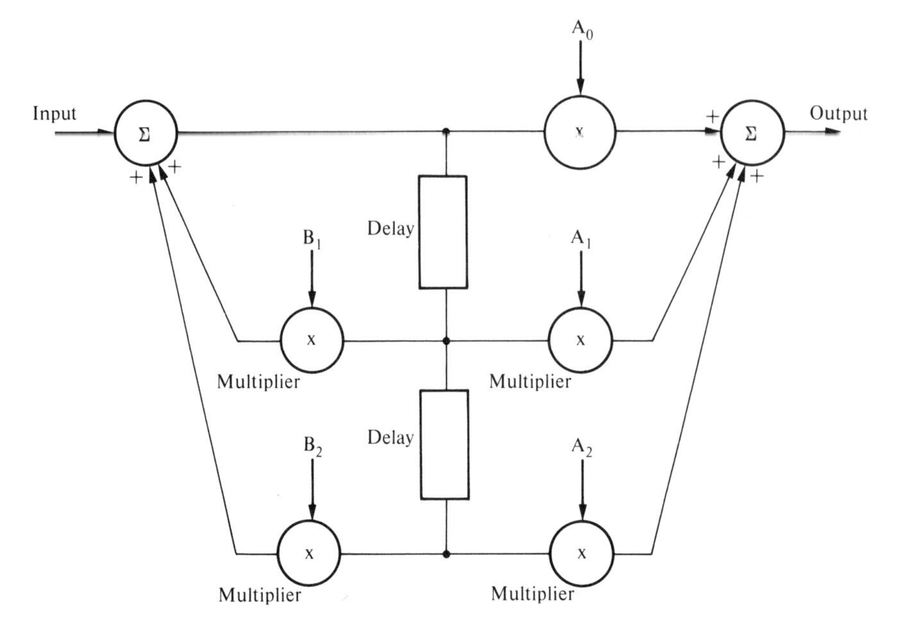

Fig. 7.14. Block diagram of a two-pole digital filter.

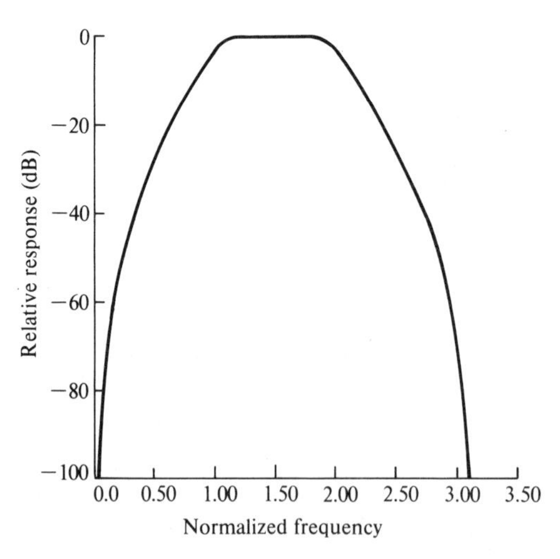

Fig. 7.15. Amplitude response of a bandpass digital filter. Normalized frequency = Frequency/(F_S/2), where F_S = sampling frequency (samples s^{-1}).

The fast Fourier transform (FFT) is an efficient method of implementing the discrete Fourier transform (DFT), which transforms a set of discrete data in the time domain into its corresponding discrete frequency spectrum and vice versa. The fundamentals and procedures of DFT and FFT will not be discussed in this section but can be found in refs. [7.13] and [7.14]. Nevertheless, the basic characteristics of FFT frequency analysers will be discussed.

The fast Fourier transform and discrete Fourier transform are based on the expressions

$$F(k) = \frac{1}{N} \sum_{n=0}^{N-1} f(n) \mathrm{e}^{-\mathrm{j}2\pi nk/N} \tag{7.1}$$

and

$$f(n) = \sum_{k=0}^{N-1} F(k) \mathrm{e}^{\mathrm{j}2\pi nk/N} \tag{7.2}$$

where $f(n)$ represents the discrete N samples of a time function $f(t)$ at N different instants evenly spaced from 0 to the record length time T, and $F(k)$ represents the discrete frequency spectrum at N different frequencies equally distributed from zero to the sampling frequency F_s.

Equations (7.1) and (7.2) are for the most general case and the samples of $f(n)$ are complex values. However, in noise measurements these are N real-value samples. These N real-value samples are to be transformed as $N/2$ complex samples so as to obtain $N/2$ complex frequency components at frequencies evenly distributed from zero to half the sampling frequency.[7.14]

In practice, a commercial FFT frequency analyser usually takes 1024 digital samples of a time function into a memory as a data block with a sampling frequency F_s. It then carries out the FFT computation, using these 1024 samples, and obtains a frequency spectrum at 512 different frequencies evenly distributed from 0 to $F_s/2$ for display on a screen. The analyser will then use the next block of 1024 samples to generate a new spectrum. The whole process from the start of FFT manipulation of the 1024 samples to the updating of the display on the screen may take much less than 0.5 second. While the analyser is performing the FFT on one data block, it is also sampling the time function and storing the data.

For clarity, Table 7.3 gives the 'basic' characteristics of a 1024-sample, 500-line (the nominal value for 512 frequencies), FFT frequency analyser. The analyser operates in 'real time', i.e. the time function is sampled at all times, when the frequency analysis range is 0 to 8 kHz or lower. For higher frequency-analysis ranges, only a fraction of the time function is sampled for the FFT manipulations and the analyser is therefore not in 'real time' operation.

TABLE 7.3. 'Basic' characteristics of a 1024-sample 500-line FFT frequency analyser (for illustration purposes)

Frequency analysis range (Hz)	Minimum† sampling frequency (S/s)	Frequency spacing (Hz)	Approximate bandwidth (Hz)	Approximate time taken to collect 1024 samples‡ (s)	Approximate time for FFT execution* (s)
0– 10	20	0.02	0.03	51	0.045
0– 100	200	0.2	0.3	5.1	0.045
0– 1 000	2 000	2	3	0.51	0.045
0– 2 000	4 000	3.9	6	0.256	0.045
0– 4 000	8 000	7.8	12	0.128	0.045
0– 8 000	16 000	15.6	24	0.064	0.045
0–16 000	32 000	31.2	48	0.032	0.045
0–32 000	64 000	62.5	96	0.016	0.045

* FFT execution time depends on the characteristics of the FFT processor hardware and software. If an array processor is used, the execution time can be much shorter than the quoted value.
† Based on the sampling theorem which states that information in a frequency band of 0–A Hz is completely defined by samples taken at a rate of at least 2A samples per second. Actual sampling frequency is usually considerably higher than the minimum.
‡ Based on the minimum sampling frequency.

Some commercial FFT frequency analysers display 400 lines in each frequency-analysis range. This is due to the fact that any low-pass filter, an essential component in an analyser, cannot completely eliminate signals at frequencies higher than a given upper cut-off frequency. This introduces errors in the frequency analysis as Shannon's sampling

TABLE 7.4. Characteristics of a 400-line FFT frequency analyser[7.19]

Frequency analysis range (Hz)	Sampling frequency (S/s)	Frequency spacing (Hz)	Time taken to collect 1024 samples (s)
0– 10	25	0.025	40
0– 20	51	0.05	20
0– 50	158	0.125	8
0– 100	256	0.25	4
0– 200	512	0.5	2
0– 500	1 280	1.25	0.8
0– 1 000	2 560	2.5	0.4
0– 2 000	5 120	5.0	0.2
0– 5 000	12 800	12.5	0.08
0–10 000	25 600	25	0.04
0–20 000	51 200	50	0.02

theorem states that a sampled time signal must not contain any components at frequencies higher than half the sampling frequency. Some analyser manufacturers therefore use and display the first valid 400 lines and discard the last 112 lines which are affected by the low-pass filter. Table 7.4 shows the characteristics of a 400-line FFT frequency analyser.

7.3.3. *Equipment for recording and analysing impulsive noise*

The noise emitted by many office and industrial machines, e.g. printers, textile machinery, pneumatic drills, and punch presses, is of an impulsive nature. It is often characterized by a high peak sound pressure of very short duration and Fig. 7.16 shows such an example for the sound-pressure time history obtained from colliding cylinders.[7.20] An impulsive sound-level meter cannot produce an accurate time history, which is useful in assessing the effects of impulsive noise on hearing damage, loudness, and annoyance and in understanding the noise generation mechanism. An accurate time history can provide us with the following information about an impulsive noise: the peak amplitude, the duration of the peak, and the repetition rate, if applicable.

In practice, an impulsive noise is often recorded first on a tape recorder or a digital event recorder and analysed later by a digital frequency analyser. Many impulsive sounds have important components at frequencies lower than 20 Hz; it is therefore necessary to use a recording system having a lower cut-off frequency down to nominally

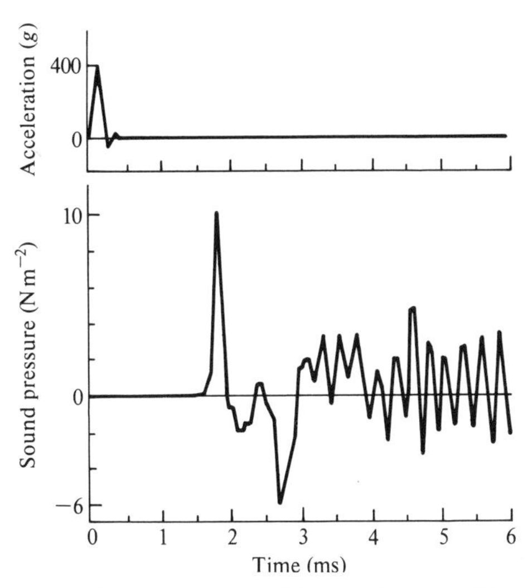

Fig. 7.16. Typical sound-pressure time history from colliding cylinders (With permission from Journal of Sound and Vibration. Copyright: Academic Press (London) Inc.).[7.20]

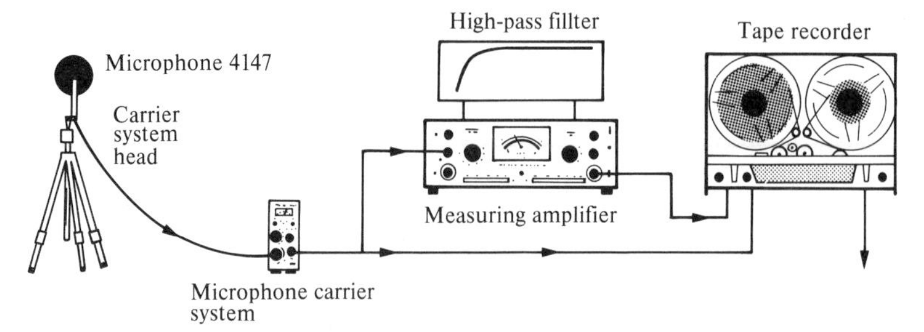

Fig. 7.17. Instrumentation system suitable for recording impulsive noise (Reproduced by permission of Brüel & Kjaer).[7.24]

zero. Furthermore, if the levels of the high-frequency components are very much lower than those of the low-frequency components, it is beneficial to split the impulsive signal into two or more frequency ranges. Figure 7.17[7.24] shows equipment suitable for recording an impulsive noise having substantial low-frequency components. The microphone and microphone-carrier system has a lower cut-off frequency of 0.01 Hz. The purpose of the additional amplifier and high-pass filter in Fig. 7.17 is to split the signal into two frequency ranges and to record it on two separate channels of the FM tape recorder. The low-level high-frequency components of the signal may be corrupted and lost by the tape noise if the high-level low-frequency components are recorded together with the high-frequency components on a single channel. It should be emphasized that a FM (frequency modulated) tape recorder is always preferred to a common direct tape recorder since the former has a nominal lower cut-off frequency of zero.

Typical analysis operations after recording the signal are

(a) Digitize the data from the tape recorder;
(b) View and plot the data to obtain the time history;
(c) Analyse selected sections to obtain the frequency spectra.

Most digital real-time frequency analysers can easily perform the above operations. However, care should be taken in selecting frequency-analysis range and weightings. For example, when using a FFT digital frequency analyser to analyse a short transient single-event impulsive noise, the transient signal should be analysed using the highest appropriate frequency range. If the transient-noise duration is 50 ms and the analyser characteristics are as shown in Table 7.3, the use of a frequency-analysis range of 0–8 kHz is advisable. Thus the transient-signal duration of 50 ms is contained within the time taken for the analyser to collect a block of 1024 samples, which is approximately 64 ms for this frequency range (see Table 7.3). If a lower frequency range of 0–4 kHz is used, the time taken to collect 1024 samples by the analyser becomes 128 ms (see Table 7.3), compared with the actual transient duration of 50 ms. This will introduce more extraneous noise in the analysis. On the other hand, if a higher frequency range of 0–16 kHz is selected, the time for collecting 1024 samples by the analyser becomes 32 ms (see Table 7.3). Thus part of the transient signal will not be sampled at all. It is therefore important to view the impulse-signal time history and to check the data-block record duration before selecting the appropriate frequency-analysis range.

Digital frequency analysers usually have either flat or Hanning weightings. For analysing a single-event, short-duration transient impulsive noise, flat weighting (i.e. no attenuation for all sampled data) is preferred to Hanning weighting. The latter is used to attenuate the sampled data at the two ends of the data block to zero, so as to eliminate discontinuities at either end of the truncation interval.

For a short, single-event, transient signal, if the signal duration is well contained in the time duration for collecting a block of 1024 samples, the two ends of the data block are already at zero value. Thus the use of Hanning weighting is not needed. If Hanning weighting were used, the important initial fluctuating data for a short transient signal would have been greatly attenuated (see Fig. 7.18).

For analysing a transient-noise signal which has components over a wide frequency range and lasts longer than the time taken to collect a block of 1024 samples for the required frequency range, a digital filter frequency analyser is preferred to a FFT frequency analyser, although special techniques can be used with a FFT analyser for analysing such a long transient signal. For these the reader should refer to refs. [7.21] and [7.11].

For an impulsive noise containing a series of impacts, it is often useful to analyse the signal at a series of consecutive time 'instants' (i.e.

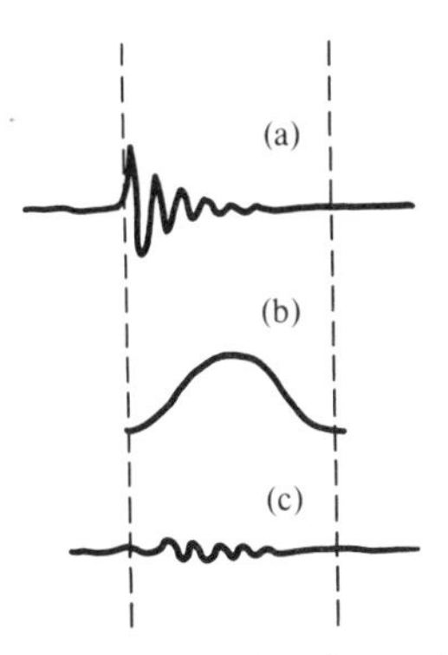

Fig. 7.18. Transient signal weighted by Hanning weighting. (a) Original transient signal; (b) Hanning weighting; (c) Weighted transient signal.

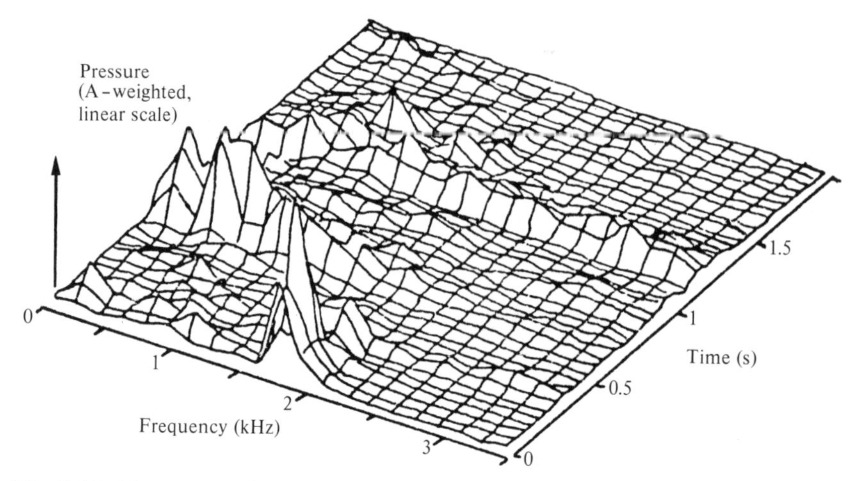

Fig. 7.19. Frequency–time–pressure diagram for the impulsive noise emitted by a punch press during one operation cycle.[7.22]

short intervals) and display the frequency spectra at different time 'instants' in a three-dimensional (3-D) frequency–time–pressure diagram. Figure 7.19 shows an example obtained by Herbert and Richard[7.22] for the impulsive noise emitted by a punch press during one cycle of its operation. The cycle consists of clutch engagement, metal piercing, and clutch release (air blow off). The noise characteristics of these three stages are clearly shown in Fig. 7.19. There are commercial digital frequency analysers incorporating facilities to perform the '3-D' analysis.

A digital frequency analyser for '2-D' analysis can also be used to produce '3-D' plots when it is assisted by a microcomputer and a digital plotter. For details the reader should refer to ref. [7.23].

7.4. Digital sound-power measurement systems

Owing to the rapid progress in microelectronics, digital noise-measurement systems to provide the sound-power levels of a machine in one-third octave bands, in octave bands, and in A-weighting are commercially available. Figure 7.20 shows one of these systems. An array of microphones placed over a predetermined measuring surface in an anechoic chamber supplies signals to a multiplexer, which is connected to a sound-power processor.

The basic operations of the sound-power processor are as follows. The signal from each microphone is first analysed by a parallel bank of one-third octave-band filters and their outputs are converted into digital signals. These digitized signals are then squared and stored in a digital

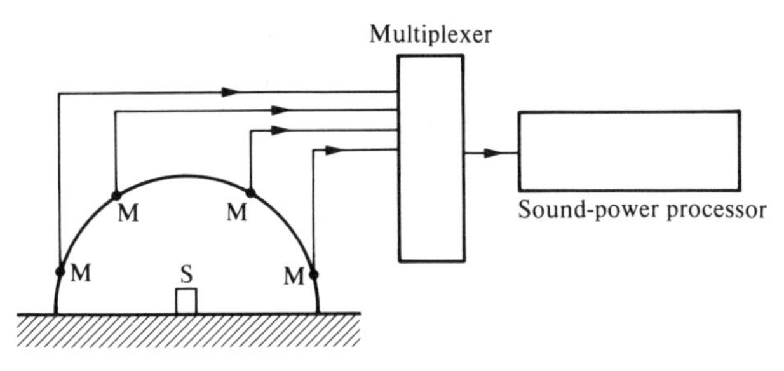

Fig. 7.20. A sound-power measurement system. M = Microphone; S = noise source.

memory. The measuring, squaring, and storing procedures are repeated for all one-third octave bands of interest, and for all microphones. The mean sound-pressure level for each one-third octave band can then be calculated, based on $\bar{L}_p = 10\log_{10}(p_{av}^2/p_{ref}^2)$. The sound-power level in each one-third octave band is obtained by adding to the mean sound-pressure level the measuring surface term, $10\log_{10}A$, based on eqn (1.9). If necessary, another correction term taking into account the effects of room pressure and temperature can also be added. By combining three one-third octave-band levels together, octave-band sound-power levels are obtained. The A-weighted sound-power level is obtained by combining all A-weighted one-third octave-band sound-power levels together.

The output of the sound-power processor can be fed in a digital form to an alphanumeric printer, a digital cassette recorder, or a visual display unit, or in an analogue form to a level recorder.

If the noise measurement is made in a reverberant room, a single microphone together with a device capable of moving the microphone slowly along a predetermined path will be sufficient for the determination of sound-power levels with the aid of the sound-power processor. The sound power level for each one-third octave band is obtained by adding to the mean sound-pressure level of the band a room-correction term based on eqn (3.31).

The cost of a digital sound-power measurement system is at present considerably higher than that of single-point analogue sound-pressure measurement equipment. However, the former offers a significant reduction in measurement time and manpower in the determination of sound power.

7.5. Dual-channel digital spectrum analysers

A dual-channel digital spectrum analyser has two input channels and can sample two signals simultaneously. In addition to the capabilities of a single-channel digital real-time frequency analyser, a dual-channel analyser can compute joint functions of the two input signals. These include cross-correlation and cross-spectral density and coherence functions, which are useful in machinery-noise analysis.

Some textbooks[7.25],[7.26] deal adequately with these functions and therefore they are only briefly discussed below with special reference to machinery-noise measurements. Let us first introduce the definition of these functions.

For two time signals $x(t)$ and $y(t)$, e.g. the signals from two microphones, the cross-correlation function between $x(t)$ and $y(t)$ is defined as

$$R_{xy}(\tau) = \lim_{T\to\infty} \frac{1}{T} \int_0^T x(t)y(t+\tau)\,dt \tag{7.3}$$

where τ is a time delay between $x(t)$ and $y(t)$. The auto-correlation function of a signal, say $x(t)$, is defined as

$$R_{xx}(\tau) = \lim_{T\to\infty} \frac{1}{T} \int_0^T x(t)x(t+\tau)\,dt. \tag{7.4}$$

The Fourier transform of $R_{xy}(\tau)$ is called the cross-spectral density function between $x(t)$ and $y(t)$, or simply the cross-spectrum between $x(t)$ and $y(t)$. Similarly, the Fourier transform of $R_{xx}(\tau)$ is called the auto-spectrum of $x(t)$ and that of $R_{yy}(\tau)$ is the auto-spectrum of $y(t)$. Mathematically, the one-sided cross-spectrum between $x(t)$ and $y(t)$ is defined as

$$G_{xy}(f) = 2\int_{-\infty}^{\infty} R_{xy}(\tau)e^{-j2\pi f\tau}\,d\tau \quad \text{for} \quad f \geqslant 0 \tag{7.5}$$

where f is frequency. The term $e^{-j2\pi f\tau}$ can be written as

$$e^{-j2\pi f\tau} = \cos(2\pi f\tau) - j\sin(2\pi f\tau). \tag{7.6}$$

Combining the eqns (7.5) and (7.6),

$$G_{xy}(f) = C_{xy}(f) - jQ_{xy}(f) \tag{7.7}$$

where

$$C_{xy}(f) = 2\int_{-\infty}^{\infty} R_{xy}(\tau)\cos(2\pi f\tau)\,d\tau \tag{7.8}$$

and

$$Q_{xy}(f) = 2\int_{-\infty}^{\infty} R_{xy}(\tau)\sin(2\pi f\tau)\,d\tau. \tag{7.9}$$

This $C_{xy}(f)$ is the real part of $G_{xy}(f)$ and is called the co-spectrum between $x(t)$ and $y(t)$ and $Q_{xy}(f)$ is the imaginary part of $G_{xy}(f)$ and is called the quad-spectrum between $x(t)$ and $y(t)$.

The magnitude of $G_{xy}(f)$ is thus given by

$$|G_{xy}(f)| = [C_{xy}^2(f) + Q_{xy}^2(f)]^{1/2}. \tag{7.10}$$

The one-sided auto-spectrum of $x(t)$ is defined as

$$G_{xx}(f) = 2\int_{-\infty}^{\infty} R_{xx}(\tau)e^{-j2\pi f\tau}\,d\tau \quad \text{for} \quad f \geqslant 0. \tag{7.11}$$

The one-sided auto-spectrum for $y(t)$ is

$$G_{yy}(f) = 2\int_{-\infty}^{\infty} R_{yy}(\tau)e^{-j2\pi f\tau}\,d\tau \quad \text{for} \quad f \geqslant 0. \tag{7.12}$$

The ordinary coherence function between $x(t)$ and $y(t)$ is defined as

$$\gamma_{xy}^2(f) = \frac{|G_{xy}(f)|^2}{G_{xx}(f)G_{yy}(f)} \tag{7.13}$$

where $|G_{xy}(f)|$, $G_{xx}(f)$, and $G_{yy}(f)$ are defined in eqns (7.10), (7.11), and (7.12), respectively.

The coherence function $\gamma_{xy}^2(f)$ is a dimensionless frequency-domain function and its value at any frequency is within 0 and +1. In machinery noise measurement, $x(t)$ could be the signal from a microphone near the surface of a machine or from an accelerometer fixed to the machine surface while $y(t)$ could be the signal from another microphone in the far field. The signal $x(t)$ can be considered as a system input and the signal $y(t)$ as a system output. If the spectral component of the output $y(t)$ at a frequency of f_i is completely correlated to the input $x(t)$ and there is no contaminating noise, the coherence function at that frequency, $\gamma_{xy}^2(f_i)$, is equal to 1. On the other hand, if the spectral component of the output at a given frequency f_j is completely uncorrelated to the input, the coherence function at that frequency $\gamma_{xy}^2(f_j)$ is zero. If the coherence function value at a frequency of f_k is neither 0 nor 1, i.e.

$$0 < \gamma_{xy}^2(f_k) < 1,$$

one or more of the following conditions exist:

(a) There is contaminating noise in the measurement;
(b) The relationship between the input and output is not linear;
(c) There are other inputs, in addition to the input $x(t)$.

In short, at a given frequency f_a, the coherence function $\gamma_{xy}^2(f_a)$ represents the fraction of the system output directly related to the system input.

A dual-channel analyser is designed to compute cross-spectrum, auto-spectrum, co-spectrum, quad-spectrum, and coherence function between two input signals for a wide frequency range. It can therefore be used to obtain the values for G_{12}, G^{s}_{12}, C_{Vp}, and Q_{Vp} in eqns (6.12) and (6.14) for measuring sound intensity. However, care should be taken about the accuracy of the results. There are random errors and bias errors in the measurement.[7.26]

In order to reduce these errors, a dual-channel FFT analyser is equipped with r.m.s. averaging. The number of r.m.s. averages is chosen as a power of 2 and varies typically from 16 to 256. The r.m.s. averaging can give a statistically more accurate estimate of the spectrum and coherence function. Table 7.5 shows the variation of the 90 per cent confidence limits on coherence and spectrum measurements with the number of r.m.s. averages.

Table 7.5 shows that for better accuracy, 32 or more averages should be used. This is particularly important for low coherence function values.

A digital spectrum analyser is also equipped with another type of

TABLE 7.5.
(a) 90 per cent confidence limits on coherence function measurements[7.27]
(Entries in tables are minimum and maximum limits)

Measured value of coherence function	Number of r.m.s. averages									
	16		32		64		128		256	
	min	max	min	max	min	max	min	max	min	max
0.4	0.15	0.59	0.23	0.54	0.28	0.50	0.32	0.47	0.34	0.45
0.5	0.25	0.67	0.33	0.63	0.39	0.59	0.42	0.57	0.45	0.55
0.6	0.36	0.74	0.45	0.71	0.50	0.68	0.53	0.66	0.55	0.64
0.7	0.50	0.81	0.57	0.78	0.61	0.76	0.64	0.75	0.66	0.73
0.8	0.65	0.88	0.70	0.86	0.74	0.84	0.76	0.83	0.77	0.82
0.9	0.81	0.94	0.85	0.93	0.87	0.92	0.88	0.92	0.88	0.91

(b) 90 per cent confidence limits on auto- and cross-spectrum measurements[7.27]

90 per cent statistical confidence limits*	Number of r.m.s. averages						
	4	8	16	32	64	128	256
Upper limit, dB	+4.7	+3.0	+2.0	+1.4	+1.0	+0.7	+0.5
Lower limit, dB	−2.9	−2.2	−1.6	−1.2	−0.8	−0.6	−0.4

* Examples. If 32 averages have been used, and the spectrum component level at a given frequency shows −50 dB, then the true level at this frequency has a 90 per cent probability of being in the range of −48.6 to −51.2 dB.

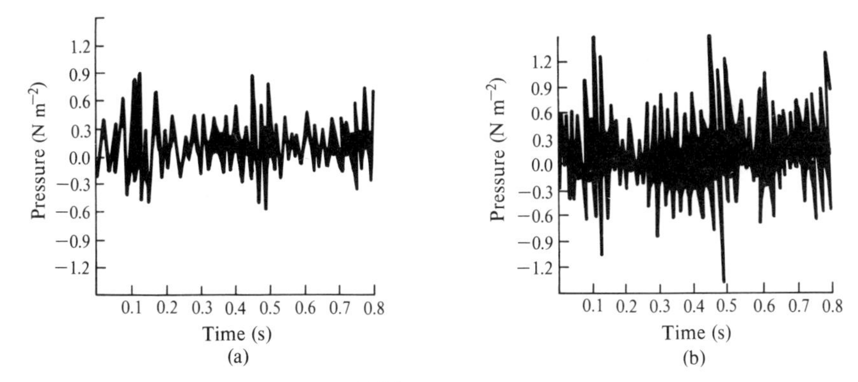

Fig. 7.21. Fly shuttle loom noise.[7.28] (a) Average noise time history ($N = 32$); (b) 'instantaneous' noise time history ($N = 1$).

averaging, time averaging, which is useful in extracting a deterministic signal from noisy background. Time averaging with N averages means that N successive, synchronized time records are added and the sum is divided by N before transformation. It can enhance the signal-to-noise ratio. For certain machines having cyclic operations, e.g. fly-shuttle looms, it is beneficial to use time averaging to obtain the noise time history. Figure 7.21[7.28] shows the time average of the loom-noise time history over a cycle after 32 averages together with the 'instantaneous', i.e. $N = 1$, loom-noise time history for the same microphone position. Only periodic components of the loom noise are retained after averaging, as the checking noise, caused by uncontrolled motion of the shuttle impact, is not noticeable in the average noise time history.

In order to identify the predominant noise-radiating parts of a machine, it is often necessary to measure the surface vibration or surface sound pressure at a number of points on various parts of the machine surface and the sound pressure in the far field. The signals from the surface vibration or surface-pressure measurements are treated as the inputs and the signal from the far-field sound-pressure measurement as the output. An example is shown in Fig. 7.22. Alfredson[7.30] placed five microphones very close to various parts of the surface of a diesel engine and their signals were taken as the inputs. The output signal was the sound pressure produced by a microphone located 1 m from the side surface. The ordinary coherence function between each of the five inputs and the output signal was determined and one of the results is shown in Fig. 7.23. For this multiple-input/single-output system it is necessary to calculate the multiple coherence function between the output and two or more of the measured inputs, defined as

$$\gamma^2_{y:x}(f) = 1 - \frac{G_{nn}(f)}{G_{yy}(f)} \tag{7.14}$$

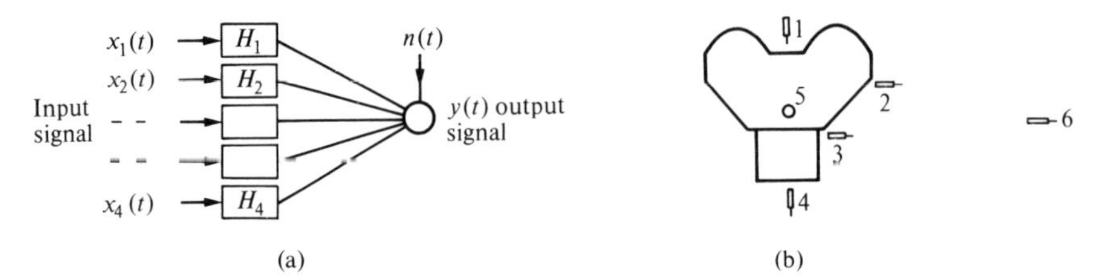

Fig. 7.22. (a) Multiple-input/single-output system. H_i = frequency response function for the ith input. (b) Locations of five input microphones (No. 1 to No. 5) and one output microphone (No. 6) for diesel engine noise measurement (With permission from Journal of Sound and Vibration. Copyright: Academic Press (London) Inc.).[7.30]

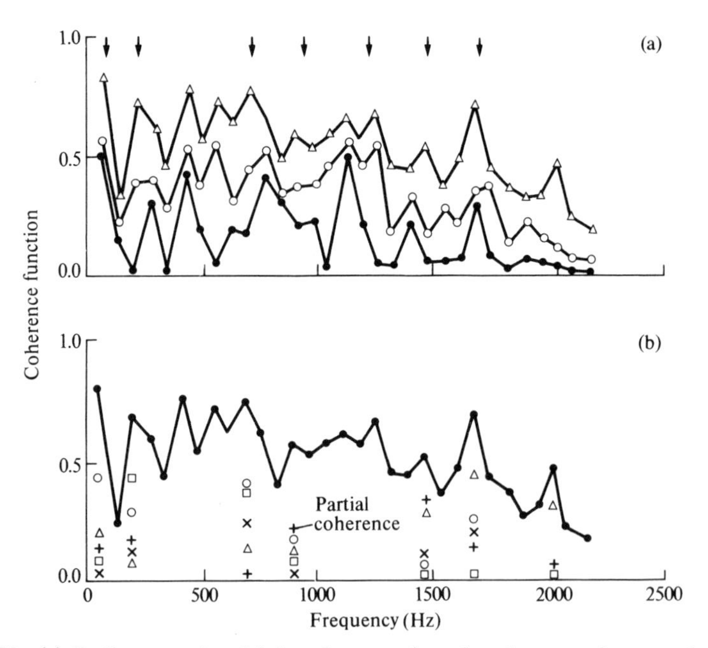

Fig. 7.23. (a) Ordinary and multiple coherence functions between input and output microphones for diesel-engine noise analysis. (With permission from Journal of Sound and Vibration. Copyright: Academic Press (London) Inc.)[7.30] (●) Ordinary coherence function between input No. 3 and output; (○) multiple coherence function between inputs (No. 1, 3, and 5) and output; (△) multiple coherence function for all five inputs and output. (Input and output locations are shown in Fig. 7.22.) (b) Multiple and partial coherence functions between input and output (+) partial coherence function for input No. 1 and output; (○) partial coherence function for input No. 2 and output (×) partial coherence function for input No. 3 and output; (△) partial coherence function for input No. 4 and output; (□) partial coherence function for input No. 5 and output.

where $G_{nn}(f)$ is the one-sided auto-spectrum of the extraneous noise signal $n(t)$ (see Fig. 7.22) and $G_{yy}(f)$ is the one-sided auto-spectrum of the output signal $y(t)$. The multiple coherence function represents the portion of the system output due to the selected two or more measured inputs. Figure 7.23 also shows the results of the multiple coherence function between three of the measured inputs (three signals from microphones 1, 3, and 5 in Fig. 7.22) and the output, and the multiple coherence function between all five measured inputs and the output. Alfredson[7.30] found that, at the significant frequencies marked by arrows in Fig. 7.23, values of the multiple coherence function between all five inputs and the output were typically of the order of 0.75. This indicated that the output, i.e. the sound pressure in the far field, was basically coherent with the five input signals, i.e. the five vibrating parts, selected in his test.

The next step is to rank the importance of these inputs. For this, the partial coherence function is helpful. For clarity, its definition is given only for a two-input/single-output system. The two inputs are $x_1(t)$ and $x_2(t)$ and the output is $y(t)$. The partial coherence function between one of the inputs, say $x_2(t)$, and the output $y(t)$ is defined as

$$\gamma^2_{2y.1}(f) = \frac{|G_{2y.1}(f)|^2}{G_{yy.1}(f)G_{22.1}(f)} \tag{7.15}$$

where $|G_{2y.1}(f)|$ is the magnitude of $G_{2y.1}(f)$, which is the cross-spectrum between $x_2(t)$ and $y(t)$ when the linear effects of $x_1(t)$ are removed from both $x_2(t)$ and $y(t)$. $G_{yy.1}(f)$ and $G_{22.1}(f)$ are the auto-spectra of $y(t)$ and $x_2(t)$, respectively, when the linear effects of $x_1(t)$ are removed from $x_2(t)$ and $y(t)$. Partial and multiple coherence functions cannot be measured directly by a dual-channel spectrum analyser. However, they can be calculated from the auto-spectra and cross-spectra which can be measured by a dual-channel analyser. For details of the calculation of partial and multiple coherence functions for multiple input/output systems, readers can refer to refs [7.26] and [7.29].

Figure 7.23 shows the partial coherence functions between each of the five inputs and the output at a number of important frequencies. It was found that no one particular input signal was highly coherent with the output signal. However, at some frequencies, one or two input signals were more highly coherent with the output than were the remaining inputs. For example, position 2 seemed more coherent with the output at 60 Hz than did the remainder. Position 4 was more coherent with the output at 1670 and 2020 Hz.

Coherence functions have also been used to study the noise sources for a punch press, the effect of cylinder pressures on diesel-engine noise emission, and the noise emitted from fly shuttle looms.

References

[7.1] Anon. (1977). *Condenser microphones and microphone preamplifiers. Theory and application handbook.* Brüel and Kjaer, Naerum.

[7.2] Frederiksen, E., Eirby, N., and Mathiasen, H. (1979). Prepolarised condenser microphones for measurement purposes. *B & K tech. Rev.* **4,** 3–26.

[7.3] IEC (1971). *Precision method for pressure calibration of one-inch standard condenser microphones by the reciprocity technique,* IEC 327. IEC, Geneva.

[7.4] IEC (1972). *Simplified method for pressure calibration of one-inch condenser microphones by the reciprocity method,* IEC 402. IEC, Geneva; IEC (1974). Precision method for free-field calibration of one-inch standard microphones by the reciprocity technique, IEC 486. IEC, Geneva.

[7.5] ANSI (1966). *Calibration of microphones,* ANSI S1.10–1966 (R 1976). Acoustical Society of America, New York.

[7.6] IEC (1973). *Precision sound level meters,* IEC 179. IEC, Geneva; IEC (1973). *First supplement to publication 179 (1973) precision sound level meters, additional characteristics for the measurement of impulsive sound,* IEC 179A. IEC, Geneva.

[7.7] BS (1967). *Specification for a precision sound level meter,* BS 4197. BSI, London.

[7.8] ANSI (1971). *Specification for sound level meters,* ANSI S1.4–1971 (R 1976). Acoustical Society of America, New York.

[7.9] IEC (1961). *Recommendations for sound level meters,* IEC 123. IEC, Geneva.

[7.10] BS (1962). *Sound level meters (industrial grade),* BS 3489. BSI, London.

[7.11] Randall, R. B. and Upton, R. (1978). Digital filters and FFT technique in real-time analysis. *B & K tech. Rev.* **1,** 3–25.

[7.12] Peled, A. and Liu, B. (1976). *Digital signal processing.* John Wiley, New York.

[7.13] Cooley, J. W. and Tukey, J. W. (1965). An algorithm for the machine calculation of complex Fourier Series. *Math. of Comp.* **19**(90), 297–301.

[7.14] Cooley, J. W., Lewis, P. A. W., and Welch, P. D. (1970). The fast Fourier transform algorithm: programming considerations in the calculation of sine, cosine and Laplace transforms. *J. Sound Vib.* **12**(3), 315–37.

[7.15] IEC (1979). *Sound level meters,* IEC 651. IEC, Geneva.

[7.16] BS (1980). *Sound level meters,* BS 5969. BSI, London.

[7.17] Martin, R. (1976). The impulse sound level meter and proposals for its use in Germany. *Proceedings of Inter-Noise 76,* pp. 117–22. Institute of Noise Control Engineering, New York.

[7.18] Brüel, P. V. (1976). Do we measure damaging noise correctly? *Proceedings of Inter-Noise 76,* pp. 111–16. Institute of Noise Control Engineering, New York.

[7.19] Broch, J. T. (1980). *Mechanical vibration and shock measurements,* 2nd edn. Brüel & Kjaer, Naerum.

[7.20] Richards, E. J., Westcott, M. E., and Jeyapalar, R. K. (1979). On the prediction of impact noise. I. Acceleration noise. *J. Sound Vib.* **62**(4), 547–75.

[7.21] Upton, R. and Randall, R. B. (1978). The application of the narrow band spectrum analyser Type 2031 to the analysis of transient and cyclic phenomena. *B & K tech. Rev.* **2,** 3–20.

[7.22] Herbert, A. G. and Richards, E. J. (1977). A new method of analysing impact noise energy. *Proceedings of Inter-Noise 77,* pp. B 343–8. International Institute of Noise Control Engineering, Zürich.

[7.23] Møller, H. (Undated). 3-Dimensional acoustic measurements—using gating techniques, *B & K Application Notes* 17–163. Brüel and Kjaer, Naerum.

[7.24] Hassall, J. R. and Zaveri, K. (1979) *Acoustic noise measurement*, 4th edn. Brüel & Kjaer, Naerum.

[7.25] Bendat, J. S. and Piersol, A. G. (1971). *Random data analysis and measurement procedures.* John Wiley, New York.

[7.26] Bendat, J. S. and Pierol, A. G. (1980). *Engineering applications of correlation and spectral analysis.* John Wiley, New York.

[7.27] Hewlett Packard (Undated). Hewlett Packard application note 245-2, Measuring the coherence function with the HP 3582A spectrum analyser; Hewlett Packard application note 245-1, Signal averaging with the HP 3582A spectrum analyser. Hewlett–Packard, Palo Alto.

[7.28] Caliskan, M., Cooke, J. A., and Bailey, J. R. (1981). Transient noise source identification in a fly-shuttle loom. *Proceedings of Noise-Con 81*, pp. 31–6. Noise Control Foundation, New York.

[7.29] Bendat, J. S. (1980). Modern analysis procedures for multiple input/output problems. *J. acoust. Soc. Am.* **68**(2), 498–503.

[7.30] Alfredson, R. J. (1977). The partial coherence technique for source identification on a diesel engine. *J. Sound Vib.* **55**(4), 487–94.

Appendix 1. Room constant and reverberation time

The room constant of a room is defined as

$$R = \frac{S\bar{\alpha}}{1-\bar{\alpha}} \tag{A.1.1}$$

where S is the surface area of the room in m^2 and $\bar{\alpha}$ is the average absorption coefficient of the room.

The reverberation time T is defined as the time taken for a sound in the room to decrease 60 dB when the sound source is suddenly stopped. The reverberation time is related to the average absorption coefficient $\bar{\alpha}$ by the expression[A1.1]

$$T = \frac{0.161V}{S\bar{\alpha}}, \tag{A.1.2}$$

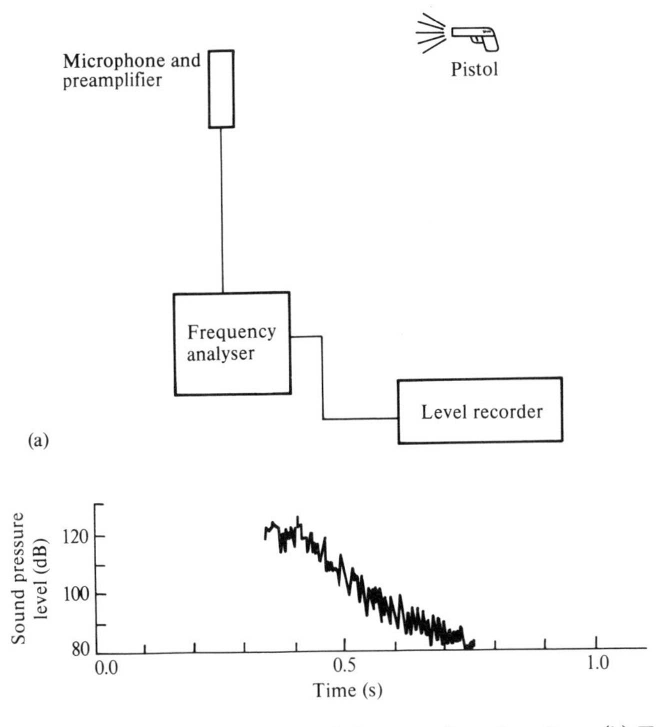

Fig. A.1.1. (a) Test arrangement for determining reverberation time. (b) Typical example of test result.

where V is the volume of the room in m^3. The reverberation time can be measured by a simple test and one test arrangement is shown in Fig. A.1.1 together with a decay curve example. Once the reverberation time is measured from the decay curve, the average absorption coefficient $\bar{\alpha}$ and the room constant R can be readily determined from eqn (A.1.2) and eqn (A.1.1). In general the reverberation time varies with frequency and should be measured for all the frequencies of interest.

Readers should refer to ISO 3382[A1.2] for details about reverberation time measurement and ref. [A1.3] for a fully automated system for measuring reverberation time using a digital frequency analyser.

References

[A1.1] Beranek, L. L. (1971). *Noise and vibration control.* McGraw-Hill, New York.

[A1.2] ISO (1975). *Measurement of reverberation time in auditoria,* ISO 3382. ISO, Geneva.

[A1.3] Upton, R. (1977). Automated measurements of reverberation time using the digital frequency analyser type 2131. *B & K tech. Rev.* **2,** 3–18.

Appendix 2. Student's t and the difference between $[(p_i^2)_{av}/p_{ref}^2]_{true}$ and $[(p_i^2)_{av}/p_{ref}^2]_{measured}$

For a set of data having a mean value of $\bar{x}_{measured}$ and a sample size of n, taken from a large population of normal (Gaussian) type having a true mean value of $\bar{x}_{true}$ and a standard deviation of σ, the following ratio is called Student's t.

$$t = \frac{|\bar{x}_{true} - \bar{x}_{measured}|}{\sigma/\sqrt{n}}.$$

Since the standard deviation σ of the large population is unknown, the following best estimate of the standard deviation for the population is used.

$$\sigma = \sigma_{sample}\sqrt{\left(\frac{n}{n-1}\right)}.$$

Combining the above two equations gives

$$|\bar{x}_{true} - \bar{x}_{measured}| = \frac{t\sigma_{sample}}{\sqrt{(n-1)}}.$$

From a set of sound-pressure level data obtained from a particular measuring surface enclosing a machine, a corresponding set can be found of p_i^2/p_{ref}^2 data and the mean value $[(p_i)_{av}^2/p_{ref}^2]_{measured}$. The relationship between the true value of p_i^2/p_{ref}^2 on the surface $[(p_i^2)_{av}/p_{ref}^2]_{true}$ and $[(p_i^2)_{av}/p_{ref}^2]_{measured}$ can thus be expressed as

$$\left|\left(\frac{(p_i^2)_{av}}{p_{ref}^2}\right)_{true} - \left(\frac{(p_i^2)_{av}}{p_{ref}^2}\right)_{measured}\right| = \frac{t\sigma_E}{\sqrt{(n-1)}}$$

where σ_E is the standard deviation for the set of p_i^2/p_{ref}^2 data and $(n-1)$ is the number of degrees of freedom. The above equation is eqn (4.1).

It can be shown[A2.1] from Student's t distribution that Student's t value will not exceed a certain critical value for a given probability and a given value of degrees of freedom, ν. Table A.2.1 gives various critical values of Student's t for a probability ranging from 10 per cent to 0.5 per cent and various numbers of degrees of freedom. For example, for a probability of 2.5 per cent and $\nu = 11$, the critical value of Student's t is found in the Table to be 2.201.

Reference

[A2.1] Weatherburn, C. E. (1962). *Mathematical statistics*. Cambridge University Press, Cambridge.

TABLE A.2.1. Critical value of Student's t

Degrees of freedom (ν)	Probability that t will be exceeded (per cent)				
	10	5	2.5	1	0.5
1	3.078	6.314	12.706	31.821	63.657
2	1.886	2.920	4.303	6.965	9.925
3	1.638	2.353	3.182	4.541	5.841
4	1.533	2.132	2.776	3.747	4.604
5	1.476	2.015	2.571	3.365	4.032
6	1.440	1.943	2.447	3.143	3.707
7	1.415	1.895	2.365	2.998	3.499
8	1.397	1.860	2.306	2.896	3.355
9	1.383	1.833	2.262	2.821	3.250
10	1.372	1.812	2.228	2.764	3.169
11	1.363	1.796	2.201	2.718	3.106
12	1.356	1.782	2.179	2.681	3.055
13	1.350	1.771	2.160	2.650	3.012
14	1.345	1.761	2.145	2.624	2.977
15	1.341	1.753	2.131	2.602	2.947
16	1.337	1.746	2.120	2.583	2.921
17	1.333	1.740	2.110	2.567	2.898
18	1.330	1.734	2.101	2.552	2.878
19	1.328	1.729	2.093	2.539	2.861
20	1.325	1.725	2.086	2.528	2.845
21	1.323	1.721	2.080	2.518	2.831
22	1.321	1.717	2.074	2.508	2.819
23	1.319	1.714	2.069	2.500	2.807
24	1.318	1.711	2.064	2.492	2.797
25	1.316	1.708	2.060	2.485	2.787
26	1.315	1.706	2.056	2.479	2.779
27	1.314	1.703	2.052	2.473	2.771
28	1.313	1.701	2.048	2.467	2.763
29	1.311	1.699	2.045	2.462	2.756
30	1.310	1.697	2.042	2.457	2.750
40	1.303	1.684	2.021	2.423	2.704
60	1.296	1.671	2.000	2.390	2.660
120	1.289	1.658	1.980	2.358	2.617
∞	1.282	1.645	1.960	2.326	2.576

Appendix 3. Derivation of the standard deviation of the total error in the mean-square sound pressure

Consider the case of a total sample size of IJK with J machines each measured at I points and with K observations (readings) taken for each measurement. From the sound-pressure level readings, there are IJK values for p_i^2/p_{ref}^2. Each of the p_i^2/p_{ref}^2 values can be represented by eqn (5.2)

$$y_{ijk} = \bar{\mu} + a_i + b_j + c_{ij} + e_{ijk}. \tag{A.3.1}$$

The averaged value of p_i^2/p_{ref}^2 based on the above equation is

$$\frac{(p_i^2)_{\text{av}}}{p_{\text{ref}}^2} = \frac{\sum_1^I \sum_1^J \sum_1^K y_{ijk}}{IJK} = \bar{\mu} + \frac{\sum_{i=1}^{I} a_i}{I} + \frac{\sum_{j=1}^{J} b_j}{J} + \frac{\sum_{i=1}^{I}\sum_{j=1}^{J} c_{ij}}{IJ} + \frac{\sum_1^I \sum_1^J \sum_1^K e_{ijk}}{IJK} \tag{A.3.2}$$

or

$$\frac{(p_i^2)_{\text{av}}}{p_{\text{ref}}^2} - \bar{\mu} = \frac{\sum a_i}{I} + \frac{\sum b_j}{J} + \frac{\sum\sum C_{ij}}{IJ} + \frac{\sum\sum\sum e_{ijk}}{IJK}. \tag{A.3.3}$$

It can be shown mathematically[A3.1] that, if

$$\chi = k_1 x_1 + k_2 x_2 + k_3 x_3 + k_4 x_4,$$

then

$$\sigma_x^2 = k_1^2 \sigma_{x_1}^2 + k_2^2 \sigma_{x_2}^2 + k_3^2 \sigma_{x_3}^2 + k_4^2 \sigma_{x_4}^2 \tag{A.3.4}$$

where k_1, k_2, k_3, and k_4 are constants. Comparing eqns (A.3.4) and (A.3.3) and noting that $k_1 = 1/I$, $k_2 = 1/J$, $k_3 = 1/IJ$, and $k_4 = 1/IJK$,

$$\mathrm{V}[(p_i^2)_{\text{av}}/p_{\text{ref}}^2 - \bar{\mu}]$$

$$= \frac{1}{I^2}\mathrm{V}\left[\sum_i a_i\right] + \frac{1}{J^2}\mathrm{V}\left[\sum_j b_j\right] + \frac{1}{K^2}\mathrm{V}\left[\sum_{i,j} c_{ij}\right] + \frac{1}{I^2J^2K^2}\mathrm{V}\left[\sum_{i,j,k} e_{ijk}\right], \tag{A.3.5}$$

where $\mathrm{V}[x] = \sigma_x^2$.

The variance of the sum of a number of independent random variates is equal to the sum of their variance.[A3.1] Thus

$$\begin{aligned} \mathrm{V}\left[\sum_{i=1}^{I} a_i\right] &= I\sigma_{\text{A}}^2, \\ \mathrm{V}\left[\sum_{j=1}^{J} b_j\right] &= J\sigma_{\text{B}}^2, \\ \mathrm{V}\left[\sum_{i=1}^{I}\sum_{j=1}^{J} c_{ij}\right] &= IJ\sigma_{\text{AB}}^2, \end{aligned} \tag{A.3.6}$$

and

$$V\left[\sum_{i=1}^{I}\sum_{j=1}^{J}\sum_{k=1}^{K} e_{ijk}\right] = IJK\sigma_e^2,$$

where σ_A^2 is the variance for factor A (points), σ_B^2 the variance for factor B (machines), σ_{AB}^2 the variance for the interaction of A and B, and σ_e^2 the variance for residual error.

Combining eqns (A.3.6) and (A.3.5) gives the variance for the total error in $(p_i^2)_{av}/p_{ref}^2$ as

$$V[(p_i^2)_{av}/p_{ref}^2 - \bar{\mu}] = \frac{\sigma_A^2}{I} + \frac{\sigma_B^2}{J} + \frac{\sigma_{AB}^2}{IJ} + \frac{\sigma_e^2}{IJK}. \qquad (A.3.7)$$

Thus the standard deviation of the total error in $(p_i^2)_{av}/p_{ref}^2$ is

$$\sigma_{tot} = \left(\frac{\sigma_A^2}{I} + \frac{\sigma_B^2}{J} + \frac{\sigma_{AB}^2}{IJ} + \frac{\sigma_e^2}{IJK}\right)^{1/2}. \qquad (A.3.8)$$

Reference

[A3.1] Brownlee, K. A. (1965). Statistical theory and methodology in science and engineering. John Wiley, New York.

Appendix 4. International and national standards on machinery noise measurements

TABLE A.4.1. International standards on machinery noise measurements*

Organization and identification of standard	Name of standard
Council for Mutual Economic Aid (COMECON)	
CT C3B 541-77	Generalized test code for the measurement of airborne noise
CT C3B 828-77	Determination of airborne noise from rotating electrical machines
CT C3B 1412-78	Determination of sound-power levels of noise sources in free-field conditions over a reflecting plane—engineering method
CT C3B 1413-78	Determination of sound-power levels of noise sources in a reverberant field—engineering method
CT C3B 1414-78	Determination of sound-power levels of noise sources—survey method
Conservation of Clean Air and Water—Europe (CONCAWE)	
CONCAWE report No. 2/76	Determination of sound power levels of noise sources in petrochemical plants
CONCAWE report No. 3/77	Test method for the measurement of noise emitted by furnaces for use in the petroleum and petrochemical industries
CONCAWE report No. 8/77	Measurement of vibrations complementary to sound measurements
CONÇAWE report No. 5/78	Method for determining the sound-power levels of air-cooled (air-fin) heat exchangers
CONCAWE report No. 2/79	Method for determining the sound-power levels of flares used in refineries, chemical plants, and oilfields
International Electrotechnical Commission (IEC)	
IEC 34-9 (1972)	Rotating electrical machines, Part 9, Noise limits
IEC 551-1976	Measurement of transformer and reactor sound levels
IEC 704-1 (1982)	Test code for the determination of airborne acoustic noise emitted by household and similar electrical appliances Part 1-General requirements

* It is intended to include all important international standards on machinery noise measurements in the table. However, it cannot be guaranteed that the table provides all these standards since new standards are adopted and existing standards may be revised or withdrawn at any time.

TABLE A.4.1. (*Continued*)

Organization and identification of standard	Name of standard
International Organization for Standardization (ISO)	
ISO R.1680-1970	Test code for the measuring of the airborne noise emitted by rotating electrical machinery
ISO/DIS 1680/1	Revision of R 1680-1970. Test code for the measurement of airborne noise emitted by rotating electrical machinery—Part 1: Engineering method for free-field conditions over a reflecting plane.
ISO/DIS 1680/2	Revision of R 1680-1970. Test code for the measurement of airborne noise emitted by rotating electrical machinery—Part 2: Survey method
ISO/DIS 3481	Measurement of airborne noise emitted by pneumatic tools and machines—engineering method for determination of sound-power levels
ISO 3740-1980	Acoustics—determination of sound-power level of noise sources—guidelines for the use of basic standards and for the preparation of noise test codes
ISO 3741-1975	Acoustics—determination of sound power sources—precision methods for broad-band sources in reverberation rooms
ISO 3742-1975	Acoustics—determination of sound power level of noise sources—precision methods for discrete-frequency and narrow-band sources in reverberation rooms
ISO 3743-1976	Determination of sound power levels of noise sources—engineering methods for special reverberation test rooms
ISO 3744-1981	Acoustics—determination of sound-power levels of noise sources—engineering methods for free-field conditions over a reflecting plane
ISO 3745-1977	Acoustics—determination of sound-power levels of noise sources—precision methods for anechoic and semi-anechoic rooms
ISO 3746-1979	Acoustics—determination of sound-power levels of noise sources—survey method
ISO/DIS 3747	Acoustics—determination of sound power levels of noise sources—survey method using a reference sound source
ISO/DIS 3748	Acoustics—determination of sound-power levels of noise sources—engineering method for small, omnidirectional sources under free-field conditions over a reflecting plane

TABLE A.4.1. (*Continued*)

Organization and identification of standard	Name of standard
ISO 3822/1-1979	Laboratory tests on noise emission by appliances and equipment used in water supply installations—Part 1: Method of measurement
ISO/DP 3822/2	Laboratory tests on noise emissions by appliances and equipment in water supply installations—Part II: Mounting and operating conditions of draw-off taps
ISO/DP 3822/3	Laboratory tests on noise emission by appliances and equipment in water supply installations—Part III: Mounting and operating conditions of in line valves and appliances
ISO/DIS 3989	Measurement of airborne noise emitted by compressor units including prime movers—engineering method for determination of sound-power level
ISO 4871-1984	Acoustics—Noise labelling of machinery and equipment
ISO 4872-1978	Measurement of airborne noise emitted by construction equipment intended for outdoor use—method for checking compliance with noise limits
ISO/DIS 5130	Measurement of noise emitted by stationary road vehicles—survey method
ISO 5131-1982	Acoustics—Tractors and machinery for agriculture and forestry—measurement of noise at the operator's position—survey method
ISO/DIS 5132	Noise emitted by earth-moving machinery—measurement at operator's work place
ISO/DIS 5133	Determination of airborne noise emitted by earth moving machinery to the surroundings—survey method
ISO 5135-1984	Determination of sound-power levels of noise from air terminal devices, high/low velocity/pressure assemblies, dampers and valves by measurement in a reverberation room
ISO/DIS 5136	Determination of sound levels of noise sources—in-duct method
ISO/DIS 5395/5	Power lawn mowers, lawn tractors with mowing attachments—safety requirements and test procedures—Part 5: Test code for the measurement of airborne noise with a view to determining compliance with noise limits

TABLE A.4.1. (*Continued*)

Organization and identification of standard	Name of standard
International Organization for Standardization (ISO)	
ISO/DIS 6081	Acoustics—noise emitted by machinery and equipment—guidelines for the preparation of test codes of engineering grade requiring noise measurements at the operator's position
ISO/DIS 6190	Method of specification and measurement of airborne noise emitted by gas turbine installations—survey method
ISO/DP 6393	Measurement of airborne noise emitted by earth-moving machinery—method for checking compliance with noise limits—exterior sound-pressure level under stationary test condition
ISO/DP 6394	Measurement of airborne noise emitted by earth-moving machinery—method for checking compliance with noise limits—operator's work-place sound-pressure level under stationary test condition
ISO/DP 6395	Measurement of airborne noise emitted by earth-moving machinery—method of determining compliance with limits for exterior noise—simulated work-cycle test conditions
ISO/DP 6396	Measurement of airborne noise emitted by earth-moving machinery—operator's position—simulated work-cycle test conditions
ISO/DP 6798	Measurement of airborne noise emitted by reciprocating internal combustion engines
Part 1	Survey method
Part 2	Engineering method
ISO/DIS 6926	Determination of sound power levels of noise sources—characterization and calibration of reference sound sources
ISO 7182-1984	Measurement at the operator's position of airborne noise from chain saws
ISO/DP 7217	Agricultural and forestry wheeled tractors and self propelled machines—measurement of noise emitted when stationary
ISO/DP 7235	Measurement procedures for ducted silencers
ISO/DP 7574	Acoustics—statistical methods for determining and verifying the noise emission values of machinery and equipment
Part 1	General
Part 2	Methods for determining and verifying labelled values for machines labelled individually
Part 3	Simple methods for determining and verifying labelled values for batches of machines

TABLE A.4.1. (*Continued*)

Organization and identification of standard	Name of standard
Part 4	Determining and verifying labelled values for batches of machines
ISO/DP 7779	Acoustics—measurement of airborne noise emitted by computer and business equipment
ISO/DIS 7960	Airborne noise emitted by machine tools
Part 1	Method for the determination of sound pressure levels at the operator's and other specified positions around the machine tool
Part 2	Operating conditions for metal working machines
Part 3	Operating conditions for wood working machines
ISO/DP 8297	Acoustics—determination of immission—relevant sound power levels of multi-source industrial plants—engineering method

TABLE A.4.2. National standards on machinery noise measurements*

Nation and identification of standard	Name of standard
Australia	
AS.1081-1975	Measurement of airborne noise emitted by rotating electrical machinery
AS.1217-1972	Methods of measurement of airborne sound emitted by machines
AS.1359	General requirements for rotating electrical machines. Pt. 51(1978): Noise level limits
AS.2012-1977	Method for measurement of airborne noise from agricultural tractor and earth-moving machinery
AS.2221	Pt. 1(1979): Engineering method for measurement of airborne sound emitted by compressor/prime mover units intended for outdoor use
	Pt. 2(1979): Engineering method for measurement of airborne sound emitted by pneumatic tools and machines
Austria	
S.5031	Determination of sound power of noise sources; method in reverberation room
S.5033	Same as ISO 3743
S.5034	Same as ISO 3744
S.5035	Same as ISO 3745
S.5036	Same as ISO 3746
Belgium	
NBN 263	Acoustic test conditions in heating, ventilation, and air-conditioning installations
S 01-200	Test codes for the measurement of airborne noise emitted by rotating electric machinery
S 01-201	Determination of acoustic power levels emitted by noise sources. Laboratory methods in reverberation chamber for broad-band sources
S 01-202	Determination of acoustic power levels emitted by noise sources. Laboratory methods in reverberation chamber for discrete-frequency and narrow-band noise sources

* It is intended to include all important national standards on machinery noise measurements and all relevant test codes from the most important professional institutions in the table. However, it cannot be guaranteed that it provides all these as new standards and test codes are adopted and existing ones may be revised or withdrawn at any time.

TABLE A.4.2. (*Continued*)

Nation and identification of standard	Name of standard
S 01-203	Determination of acoustic-power levels emitted by noise sources. Survey methods for special test reverberation chambers
S 01-205	Determination of acoustic-power levels emitted by noise sources. Laboratory methods for anechoic and semi-anechoic chambers
S 01-206	Determination of acoustic-power levels emitted by noise sources. Control method
S 01-403	Noise produced by hydraulic equipment
Bulgaria	
BDS 6011-66	Measurement of noise emitted by rotating electrical machines
Czechoslovakia	
ČSN 09 0862	Noise of diesel engines. Method of measurement
ČSN 12 3062	Fans. Prescriptions for measurement of noise
ČSN 17 8055	Measurement of noise emitted by computers
ČSN 35 0000	Measurement of noise emitted by electrical machines
ČSN 35 0019	Special testing methods for rotating electrical machines. III. Noise measurement
ČSN 35 1080 1968	Fundamental tests of power transformers and reactors
ČSN 36 1005	Noise measurement of domestic electrical motor-operated appliances
ČSN 36 1006 1969	Measurement of noise emitted by large electrical household appliances
Denmark	
DS/ISO/R495 1969	Same as ISO R495
DS/ISO/3741 1975	Same as ISO 3741
DS/ISO/3742 1975	Same as ISO 3742
DS/ISO/3743 1979	Same as ISO 3743
DS/ISO/3744 1982	Same as ISO 3744
DS/ISO/3745 1979	Same as ISO 3745
DS/ISO 3746 1980	Same as ISO 3746
DS/ISO/4872 1978	Same as ISO 4872
France	
S 30-006 1976	Determination of sound power levels of noise sources. Part O—Guidelines for the use of basic standards and for the preparation of noise test codes

TABLE A.4.2. (*Continued*)

Nation and identification of standard	Name of standard
S 31-006 1979	Test codes for measuring noise emitted by rotating electric machinery—method of assessment in free field over a reflecting plane
S 31-020 1975	Test codes for measuring airborne noise emitted by motor-compressors
S 31-021 1977	Measurement of noise emitted by ducted fans
S 31-022 1973	Determination of acoustic power emitted by noise sources—Part 1—laboratory method in a reverberation chamber for broad-band small sources
S 31-023 1973	Determination of acoustic power emitted by noise sources—Part 2—laboratory method in a reverberation chamber for discrete frequencies or narrow-band small sources
S 31-024 1973	Determination of acoustic power emitted by noise sources—Part 3—engineering method for special reverberation chambers
S 31-025 1977	Determination of acoustic power emitted by noise sources—Part 4—engineering method for free-field conditions over a reflecting plane
S 31-026 1978	Determination of acoustic power emitted by noise sources—Part 5—laboratory method in anechoic and semi-anechoic rooms
S 31-027 1977	Determination of acoustic power emitted by noise Sources—Part 6—Control method for on-site measurements
S 31-030 1977	Test codes for measuring airborne noise emitted by hydraulic, electric and thermic pick hammers and concrete-breakers
S 31-031 1975	Test codes for measuring airborne noise emitted by pneumatic machine tools
S 31-060 1977	Test codes for measuring airborne noise emitted by textile machinery—survey and engineering methods
S 31-032 1974	Measurement of noise emitted by earthwork and hoisting equipment—general requirements for test codes
S 31-033 1974	Test codes for measuring noise emitted by mechanical and hydraulic shovels—operating conditions and test point locations
S 31-034 1974	Test codes for measuring noise emitted by chargers on pneumatic machines—operating conditions and test point locations
S 31-035 1974	Test codes for measuring noise emitted by fixed concrete-mixers—operating conditions and test point locations
S 31-036 1974	Test codes for measuring noise emitted by portable concrete-mixers—operating conditions and test point locations

TABLE A.4.2. (*Continued*)

Nation and identification of standard	Name of standard
S 31-037 1974	Test codes for measuring noise emitted by electrical generating groups—operating conditions and test point locations
S 31-038 1974	Test codes for measuring noise emitted by perforating machines—operating conditions and test point locations
S 31-039 1974	Test codes for measuring noise emitted by loaders on chains—operating conditions and test point locations
S 31-041 1975	Measurement of noise at the working position of the operator of agricultural tractors and machinery
S 31-042 1974	Test codes for measuring airborne noise emitted by vibrating rams
S 31-043 1975	Test codes for measuring airborne noise emitted by vibrating rollers
S 31-044 1975	Test codes for measuring airborne noise emitted by vibrating blocks and engines with vibrating pads
S 31-058 1976	Survey method for the measurement of noise emitted by stationary road vehicles
S 31-061 1978	Machinery intended for open air use. Guidelines for the preparation of the test codes for the verification of conformity to noise limits
Germany, Federal Republic of	
DIN 42540	Audible sound levels for transformers, noise power
DIN 45402	Measurement of the r.m.s. value in sound system equipment; test method for measuring devices
DIN 45635	Measurement of airborne noise emitted by machines
Note 1	Enveloping surface method, form for measurement report (1979.02)
Note 2	Explanatory remarks on the noise emission quantities (1977.12)
Note 3	List of machines dealt with in DIN 45635-Parts (1982.10)
Part 1	Enveloping surface method (1972.01)
E Part 1	Basic requirements for three accuracy classes (1982.01)
Part 2	Reverberation room method, basic measurement method (precision method grade 1) (1977.12)
Part 3	Engineering method for special reverberation test rooms (1978.09)
Part 8	Measurement of structure-borne noise, basic requirements (1981.07)
Part 9	In-duct method (1977.09)
Part 10	Enveloping surface method—rotating electrical machines (1974.05)

TABLE A.4.2. (*Continued*)

Nation and identification of standard	Name of standard
Part 11	Enveloping surface method—reciprocating internal combustion engines (1974.09)
Part 12	Enveloping surface method—electrical switchgear and control gear (1978.03)
Part 13	Enveloping surface method—compressors, vacuum pumps (displacement, turbo and jet compressors) (1977.02)
Part 14	Enveloping surface method—air cooled heat exchangers (1980.07)
Part 15	Enveloping surface method—turbine sets (1976.02)
Part 16	Enveloping surface method—machine tools (1978.06)
Part 17	Enveloping surface method—portable chain saws driven by combustion engines (1978.03)
Part 18	Enveloping surface method—appliances for household and similar purposes (1976.01)
Part 19	Enveloping surface method—office machines (1978.08)
Part 20	Enveloping surface method—pneumatic tools and machines (1975.07)
Part 21	Enveloping surface method—electric tools (1977.12)
Part 22	Enveloping surface method—flares (1983.07)
Part 23	Enveloping surface method—gear transmissions (1978.07)
Part 24	Enveloping surface method—liquid pumps (1980.03)
Part 25	Enveloping surface method—oxy-fuel gas and plasma torches and machines (1980.11)
Part 26	Enveloping surface method—hydraulic pumps (1979.07)
Part 27	Enveloping surface method—printing and paper processing machines (1978.09)
Part 28	Enveloping surface method—packaging and allied machinery (1980.11)
Part 29	Enveloping surface method—food, drink and allied machinery (1980.11)
Part 30	Enveloping surface method—transformers and reactors (1981.04)
Part 31	Enveloping surface method—pulverizing machines (1980.11)
Part 31/Note 1	Enveloping surface method—form for measurement report on pulverizing machines (1979.02)
Part 32	Enveloping surface method—textile machines (1980.05)
Part 32/Note 1	Enveloping surface method—form for measurement report on textile machines (1981.04)

TABLE A.4.2. (*Continued*)

Nation and identification of standard	Name of standard
Part 33	Enveloping surface method—construction equipment (1979.07)
Part 34	Enveloping surface method—cartridge-operated fixing tools (1977.08)
Part 35	Enveloping surface method—heat pumps (1983.12)
Part 36	Enveloping surface method—sit-on fork-lift trucks powered by internal combustion engines (1981.03)
Part 36/Note 1	Enveloping surface method—form for measurement report on sit-on fork-lift trucks powered by internal combustion engines (1981.03)
Part 37	Enveloping surface method—machinery for the processing of plastics and rubber (1980.11)
Part 38	Enveloping surface method—fans (1980.06)
Part 39	Enveloping surface method—processing furnaces (1983.08)
Part 40	Enveloping surface method—machines for hydroelectric power stations and water-pumping plants (1981.07)
Part 41	Enveloping surface method—hydraulic assemblies (1982.01)
Part 41/Note 1	Enveloping surface method—form for measurement report on hydraulic assemblies (1982.01)
Part 42	Enveloping surface method—machines of pulp and paper industry (1982.01)
Part 44	Enveloping surface method—refuse collection vehicle (1982.08)
Part 45	Enveloping surface method—continuous handling equipment (1982.07)
Part 46	Enveloping surface method—cooling towers (1983.07)
Part 47	Enveloping surface method—chimneys (1983.07)
Part 49	Enveloping surface method—surface treatment plant (1983.05)
Part 49/Note 1	Enveloping surface method—form for measurement report on surface treatment plant (1983.05)
Part 202	Enveloping surface method—machine tools (modification and supplement to 45635, Part 18) (1978.12)
Part 1601	Enveloping surface method—metal working machines, lathes (1978.07)
Part 1602	Enveloping surface method—metal working machines, drop forging hammers (1978.06)
Part 1603	Enveloping surface method—metal working machines, universal presses (1978.06)
Part 1605	Enveloping surface method—metal working machines, milling machines (1981.04)

TABLE A.4.2. (*Continued*)

Nation and identification of standard	Name of standard
Part 1606	Enveloping surface method—metal working machines, drilling machines (1981.01)
Part 1607	Enveloping surface method—metal working machines, gear-hobbing machines (1981.01)
Part 1609	Enveloping surface method—metal working machines, circular saws (1981.01)
Part 1610	Enveloping surface method—metal working machines, grinding machines (1981.01)
Part 1650	Enveloping surface method—woodworking machines, planing machines (1978.07)
Part 1651	Enveloping surface method—woodworking machines, single-blade circular saw benches (1978.06)
Part 1652	Enveloping surface method—woodworking machines, one-side moulding machines (1978.06)
Part 1653	Enveloping surface method—woodworking machines, double-end profiling machines (1982.10)
Part 1654	Enveloping surface method—woodworking machines, multistage edge, lipping and banding machines (1982.10)
Part 1655	Enveloping surface method—woodworking machines, final trimming, edge lipping and banding machines (1982.10)
Part 1656	Enveloping surface method—woodworking machines, two-sided and multi-sided planing and milling machines (1982.10)
Part 1657	Enveloping surface method—woodworking machines, double-dimensional circular sawing machines (1983.08)
Part 1658	Enveloping surface method-woodworking machines, single blade stroke circular sawing machines for cross-cutting (1983.07)
DIN 45636	Measurement of noise emitted by motor vehicles (1967)
DIN 45639	Inside noise of motor vehicles (1969)
DIN 45649	Acoustics—noise labelling of machinery and equipment (1982.07)
DIN 45650	Acoustics—statistical methods for verifying noise emission values of machinery and equipment (1982.07)
DIN IEC 59F(CO)27	Test code for the determination of airborne noise emitted by household and similar electrical appliances—Part 2; particular requirements for vacuum cleaners (1982.11)

TABLE A.4.2. (*Continued*)

Nation and identification of standard	Name of standard
Germany, Democratic Republic of	
TGL 39-440	Measurement of noise emitted by the gear box of motor vehicles
TGL 45-01248	Measurement of noise; sound level of sewing machines
TGL 50-29034	Measurement of noise emitted by rotating electrical machines; guidelines
TGL 153-6011	Measurement of noise emitted by bearings
TGL 152-6012	Explanatory remarks to noise emission quantities of different bearings
TGL 21814 6173	Measurement of airborne noise emitted by gear box (gear transmission) Type 10 LA (transmission ratio = 6.3 to 40)
TGL 21815 6173	Measurement of airborne noise emitted by gear box (gear transmission) Type 10 LA (transmission ratio = 40 to 250)
TGL 27641 BL.8	Testing of transformers, measurement of airborne noise
TGL 200-3110	Electrical machines; difinitions; procedure for calculating noise levels
TGL 200-4504	Appliances for household and similar purposes; measurement of airborne noise
TGL 22423-2 1969	Rotating electrical machines; test procedures for measurement of noise
Hungary	
MSZ KGST 380-81	Determination of sound-power levels of noise sources. Precision method for special reverberation test rooms
MSZ KGST 541-77	Noise measurement. Methods for measurement. General specifications
MSZ KGST 828-77	Rotating electrical machines. Noise measuring methods
MSZ KGST 1348-78	Rotating electrical machines. Permissible noise limits
MSZ KGST 1412-78	Determination of sound-power levels of noise sources. Engineering method for free-field conditions over a reflecting plane
MSZ KGST 1413-78	Determination of sound-power levels of noise sources. Survey methods
MSZ KGST 1414-78	Determination of sound-power levels of noise sources. Engineering methods for special reverberation test rooms
MSZ KGST 3076-81	Determination of sound-power levels of noise sources. Precision method for special anechoic acoustical test rooms
MSZ 3392-54	Acoustical measurements

TABLE A.4.2. (*Continued*)

Nation and identification of standard	Name of standard
India	
IS: 4758-1968	Methods of measurement of noise emitted by machines
IS: 6098-1971	Methods of measurement of the airborne noise emitted by rotating electrical machinery
Japan	
JIS B 1548 (1976)	Measuring methods of sound-pressure levels of ball- and roller-bearings
JISB 1753 (1976)	Measuring method of noise of gears
JIS B 6004 (1962)	Method of sound-level measurement for machine tools
JIS B 8005 (1975)	Measuring method of noise emitted by internal combustion engines
JIS B 8310 (1977)	Methods of noise level measurement for pumps
JIS B 8346 (1977)	Methods of noise level measurement for fans, blowers and compressors
JIS B 8350 (1977)	Methods of noise level measurement for oil hydraulic pumps
JIS D 1024 (1976)	Measurement of noise emitted by automobiles
JEC 37 (1979)	Induction machines. 12.5, noise level measurements
Netherlands	
NEN 21680	Same as ISO R 1680-1970
Norway	
NS 4808	General requirements for the preparation of test codes for measuring the noise emitted by machines
Poland	
PN-81 E-04257	Rotating electrical machinery. Determination of acoustic noise level
PN-75 E-06260	Appliances for domestic and similar purposes. Noise level. Examinations and principles of fixing of admissible level
PN-72 M-43120	Fans. Methods of noise determination
PN-75 M-47015	Earth moving machinery. Operator's stand. Admissible noise level and methods of tests
PN-77 M-55725	Machine tools for metals. Test methods and admissible noise levels

TABLE A.4.2. (*Continued*)

Nation and identification of standard	Name of standard
PN-75 M-78080	Driven carriageway cars. Admissible noise level and methods of tests
PN-71 N-01300	Noise of machines and equipment. Methods for determination of acoustic parameters
PN-76 E-04072	Transformers; determination of parameters of the noise
PN-76 R-36125	Agriculture tractors and machinery. Noise level at the operator's workplace. Measurement method.
Romania	
STAS 1703/6-77	Power transformers, methods of sound level measurement
STAS 7150-77	Methods of noise measurement in industry
STAS 7301-74	Measurement of noise emitted by rotating electrical machines
STAS 10834-77	Radial fans, acoustic testing methods
South Africa	
SABS 063-1972	The measurement under acoustic laboratory conditions of sound emitted by sound sources (determination of sound power levels under laboratory conditions)
SABS 069-1972	The measurement in the field of sound emitted by sound sources (determination of sound-power levels under practical operation conditions)
SABS 097-1975	The measurement of noise emitted by motor vehicles
Spain	
20 121 75	Acceptable noise levels for rotating electrical machinery
20 136 78	General requirements for the preparation of test codes for measuring the noise emitted by machines
20 137 78	Test codes for measuring airborne noise emitted by rotating electrical machinery
UNE 26231	Method of measurement of the noise emitted by vehicles
Sweden	
SEN 330501	Cooling fan; determination of noise level capacity
SMS/ISO R2151	Measurement of airborne noise emitted from motor-driven outdoor compressors

TABLE A.4.2. (*Continued*)

Nation and identification of standard	Name of standard
SMS 965/SS 3176-3179	Measurement of noise; calculation of sound-power level of woodworking machines/data sheets
SMS 892	Measurement of noise emission to the environment by machines used in building construction sites (prepared in 1973)
SMS 2189	Measurement of noise from chain saws
BAS	Test for noise level emitted from rotating grinding, milling, and drilling machines
SISO 25131	Measurement of noise emitted by motor vehicles
United Kingdom	
BEAMA Publ. No. 227	Guide to transformer noise measurement
BEAMA Publ. No. 225	BEAMA recommendations for the measurement and classification of acoustic noise from rotating electrical machines
BEBS T1, T3	Specifications for transformers
BS 848: Part 2: 1966	Fan noise testing
BS 3425: 1966	Method for the measurement of noise emitted by motor vehicles
BS 4196: Part 0: 1981	Guide for the use of basic standards and for the preparation of noise test codes
BS 4196: Part 1: 1981	Precision methods for determination of sound power levels for broad-band sources in reverberation rooms
BS 4196: Part 2: 1981	Precision methods for determination of sound-power levels for discrete-frequency and narrow-band sources in reverberation rooms
BS 4196: Part 3: 1981	Engineering methods for determination of sound-power levels for sources in special reverberation test rooms
BS 4196: Part 4: 1981	Engineering methods for determination of sound-power levels for sources in free-field conditions over a reflective plane
BS 4196: Part 5: 1981	Precision methods for determination of sound-power levels in anechoic and semi-anechoic rooms
BS 4196: Part 6: 1981	Survey method for determination of sound-power levels of noise sources
BS 4718: 1971	Methods of test for silencers for air distribution systems
BS 4773: Part 2: 1976	Methods of testing and rating air terminal devices for air distribution systems, acoustic testing
BS 4813: 1972	Method of measuring noise from machine tools excluding testing in anechoic chambers

TABLE A.4.2. (*Continued*)

Nation and identification of standard	Name of standard
BS 4856: Part 4: 1978 Part 5: 1979	Methods for testing and rating fan coil units—unit heaters and unit coolers. Part 4: Acoustic performance: without additional ducting; Part 5: Acoustic performance with ducting
BS 4857: Part 2: 1978	Methods for testing and rating terminal reheat units for air distribution systems, acoustic testing and rating
BS 4954: Part 2: 1978	Method for testing and rating induction units for air distribution systems, acoustic testing and rating
BS 4999: Part 51: 1973 (Amend. 1977 and 1978)	General requirements for rotating electrical machines; noise levels
BS 5944: Part 4: 1984	Measurement of airborne noise from hydraulic fluid power systems and components—method of determining sound power levels from valves controlling flow and pressure
OCMA Specification No. NWG 1 (Rev. 2) 1980	Noise procedure specification
OCMA Publication No. NWG 3 (Rev. 2) 1980	Guide to the use of noise procedure specification NWG 1
OCMA Specification No. NWG 4 1980	General specification for silencers and acoustic enclosures
USA	
AFBMA 13 (1970 (R 1977))	Rolling bearing vibration and noise
AGMA 295.03 (1977)	Specification for measurement of sound on high-speed helical and herringbone gear units
AGMA 297.01 (1973)	Sound for enclosed helical, herringbone and spiral-bevel gear drives
AGMA 298.01 (1975)	Sound for gear motors and in-line reducers and increasers
ANSI S1.21-1972	Methods for the determination of sound-power levels of small sources in reverberation rooms
ANSI S1.23-1976	Method for the designation of sound power emitted by machinery and equipment
ANSI S1.29-1979	Method for the measurement and rating of noise emitted by computer and business equipment
ANSI S1.30-1979	Guidelines for the use of sound-power standards and for the preparation of noise test codes
ANSI S1.31-1980	Precision methods for the determination of sound-

TABLE A.4.2. (*Continued*)

Nation and identification of standard	Name of standard
	power levels of broadband noise sources in reverberation rooms
ANSI S1.32-1980	Precision methods for the determination of sound-power levels of discrete-frequency and narrow-band noise sources in reverberation rooms
ANSI S1.33-1982	Engineering methods for the determination of sound power levels of noise sources in a special reverberation test room
ANSI S1.34-1980	Engineering methods for the determination of sound-power levels of noise sources for essentially free-field conditions over a reflecting plane
ANSI S1.35 1979	Precision methods for the determination of sound-power levels of noise sources in anechoic and hemi-anechoic rooms
ANSI S1.36-1979	Survey methods for the determination of sound-power levels of noise sources
ANSI S1.39 (Draft)	Guidelines for the preparation of standard procedures for the measurement of source sound emission
ANSI S3.17-1975	Method for rating the sound-power spectra of small stationary noise sources
ANSI S5.1-1971	Test code for the measurement of sound from pneumatic equipment
ANSI S 12.1-1983	Guidelines for the preparation of standard procedures to determine the noise emission from sources
ARI Standard 575 (1973)	Standard for method of measuring machine sound within equipment rooms
ASA STD 3 1975	Test-site measurement of noise emitted by engine powered equipment
ASA STD 4 1975	Method for rating the sound power spectra of small stationary noise sources
ASA STD 5 1976	Method for the designation of sound power emitted by machinery and equipment
ASHRAE 36-72	Methods of testing for sound rating heating, refrigerating, and air-conditioning equipment
ASHRAE 68-78	Method of testing in duct sound power measurement procedure for fans
ATMA Test Procedure (1973)	Noise measurement technique for textile machinery
Compressed Air & Gas Institute (1969)	Test code for the measurement of sound from pneumatic equipment
Diesel Engine Manufacturers Association 1972	DEMA test code for the measurement of sound from heavy-duty reciprocating engine

TABLE A.4.2. (*Continued*)

Nation and identification of standard	Name of standard
Home Ventilating Institute 1968	HVI Test procedure. Sound test procedure
IEEE Std 85 1973	Test procedure for airborne sound measurements on rotating electric machinery
NEMA Standard TR1-1974	Transformers, regulators, and reactors. Section 9-04: Audible sound level tests
NFPA T3.9.70.12 1970 (R 1975)	Method for measuring sound generated by hydraulic fluid power pumps
NFPA T3.9.14 1971 (R 1976)	Method for measuring sound generated by hydraulic fluid power motors
National Machine Tool Builders Association	Noise measurement techniques
SAE J1046 (1976)	SAE recommended practice, exterior sound-level measurement procedure for small engine-powered equipment
SAE J1074 (1974)	SAE recommended practice, engine sound-level measurement procedure
SAE J366b (1973)	Exterior sound level for heavy trucks and busses
SAE J986b (1979)	Sound level for passenger cars and light trucks
SAE J1174 (1977)	SAE recommended practice, operator ear sound level measurement procedure for small engine powered equipment
WMMA Test Code (1973)	Test code for evaluating the noise emission of woodworking machinery
USSR	
Gost 11929-66	Measurement of noise emitted by rotating electrical machines and transformers
Gost 8.055-73	State system for ensuring the uniformity of measuring. Machines—measurement methods for the determination of noise characteristics
Gost 15529-70	Ventilators for general purposes. Methods for determination of noise characteristics
Gost 16317-70	Household refrigerators
Gost 20445-75	Building and structures of industrial enterprises. Method of noise measurements at workplaces
Gost 23789-79	Noise. Methods of the measurement of noise reduction of ventilating, air conditioning, and hot-air system silencers
Gost 12.1.023-80	System of standards for labour safety. Noise—determination methods of stationary machine noise characteristic values

Index